# INTERFACIAL SCIENCE

# INTERFACIAL SCIENCE

## AN INTRODUCTION

**Geoffrey Barnes**

**Ian Gentle**

The University of Queensland

OXFORD

UNIVERSITY PRESS

# OXFORD
UNIVERSITY PRESS

Great Clarendon Street, Oxford OX2 6DP

Oxford University Press is a department of the University of Oxford.
It furthers the University's objective of excellence in research, scholarship,
and education by publishing worldwide in

Oxford New York

Auckland Cape Town Dar es Salaam Hong Kong Karachi
Kuala Lumpur Madrid Melbourne Mexico City Nairobi
New Delhi Shanghai Taipei Toronto

With offices in

Argentina Austria Brazil Chile Czech Republic France Greece
Guatemala Hungary Italy Japan Poland Portugal Singapore
South Korea Switzerland Thailand Turkey Ukraine Vietnam

Oxford is a registered trade mark of Oxford University Press
in the UK and in certain other countries

Published in the United States
by Oxford University Press Inc., New York

British Library Cataloguing in Publication Data

Data available

Library of Congress Cataloging in Publication Data

Gentle, Ian.
An introduction to interfacial science / Ian Gentle, Geoffrey Barnes.
p. cm.
ISBN–13: 978-0-19-927882-4
ISBN–10: 0-19-927882-2
1. Interfaces (Physical sciences) 2. Surface chemistry. I. Barnes, Geoff. II. Title.
QC173.4.I57G48 2005
530.4′17—dc22                                                    2005011113

Typeset by Newgen Imaging Systems (P) Ltd., Chennai, India
Printed in Great Britain
on acid-free paper by
Ashford Colour Press Ltd., Gosport, Hants.

ISBN 0-19-927882-2 (Pbk.)

1 3 5 7 9 10 8 6 4 2

# Preface

Although the history of the study of interfaces extends well over a century, its relevance is increasing rather than diminishing due to the emergence of areas of science that we may loosely describe by the term *nanoscience*. As the basic functional and structural unit becomes smaller and approaches molecular dimensions, the influence of the interface cannot be ignored.

It is therefore important that students of chemistry, engineering and materials science should have some understanding of interfacial science. Most textbooks of physical chemistry include a chapter on interfaces or surfaces, but often this does little more than alert the student to their existence and provide a very brief outline of the subject.

There are, of course, specialist texts dealing with the science of interfaces and surfaces. They include substantial and authoritative volumes on selected aspects and we have referred to these extensively throughout this book. For comprehensive texts covering the whole subject it would not be correct to state that there is a full range available as we have found that there is a gap between the advanced texts at one extreme and smaller texts providing a fairly qualitative treatment at the other. This gap occurs at the level at which we would like to introduce the subject to undergraduate students, so it is the lack of a suitable text that has provided the motivation for writing the current volume.

This book is intended for intermediate and senior undergraduate students in chemistry and related subjects but should also serve as an introduction to the topic for research students in this and other fields. For the student, we have assumed some knowledge of physical chemistry and of thermodynamics in particular, but even without this background the book can be read and hopefully understood, as the more complex derivations are shaded so that they can be skipped without losing the thread of the argument. For the instructor and the more advanced student, we have provided references to additional reading and limited but not exhaustive references to original research articles.

In writing an introductory text, we have endeavoured to provide an overview of the whole field with sufficient detail to allow an understanding of the material in both a qualitative and a quantitative sense. The emphasis is on the basic science as it is understood in the twenty-first century. The latest research is included only when it impacts on fundamental concepts or as boxed inserts when it has been judged to be of special interest. One important topic has been excluded: solid–solid interfaces are the basic features in most electronic devices and have been the subject of an enormous amount of research and development in recent years. Consequently the literature is immense and we did not think it possible to treat it adequately in a short chapter of this book.

We are greatly indebted to past and present colleagues and students for various contributions: data for figures, diagrams, images, and most of all for tolerance and understanding. They are: Geoff Ashwell, Bronwyn Battersby, Ian Costin, David Gorwyn, Rob Lamb, Gwen Lawrie, Marylyn McGregor, Jian-Bang Peng, Jeremy Ruggles, Brendan Sinnamon, Roland Steitz and Barry Wood.

We would also like to thank Bob Hunter for taking the time to read some of the text and offer detailed comments. His contribution has been most helpful.

Finally, for tolerance, patience, and understanding, we would like to give special thanks to our wives: Beth and Jillian.

Geoff Barnes
Ian Gentle
Brisbane, 2005

# Contents

# Acknowledgements

We acknowledge with thanks permission to use copyrighted material as listed in the following table.

Figure 2.20    Data from N. B. Vargaftik, B. N. Volkov, and L. D.Voljak (1983). International tables of the surface tension of water. *J. Phys. Chem. Ref. Data*, **12**, 817. Copyright 1983, with permission from the American Institute of Physics.

Figure 2.21    Adapted from F. H. Stillinger and A. Rahman (1974). Improved simulation of liquid water by molecular dynamics. *J. Chem. Phys.*, **60**, 1545. Copyright 1974, with permission from the American Institute of Physics.

Figure 4.3     Data from A. M. Posner, J. R. Anderson and A. E. Alexander (1952). The surface tension and surface potential of aqueous solutions of normal aliphatic alcohols. *J. Colloid Sci.*, **7**, 623. Copyright 1952, with permission from Elsevier.

Figure 4.7     R. K. Schofield and E. K. Rideal (1925). The kinetic theory of surface films. Part I. *Proc. Roy. Soc. A*, **109**, 57. Replotted from Figure 3 with permission from The Royal Society (London).

Figure 4.15    Replotted with permission from W. C. Preston (1948). Some correlating principles of detergent action. *J. Phys. Colloid Chem.*, **52**, 84. Copyright 1948 American Chemical Society.

Figure 4.17    Reprinted from J. N. Israelachvili, *Intermolecular and Surface Forces, 2nd edn.*, Fig. 17.3. Copyright 1991, with permission from Elsevier.

Figure 4.18    Reprinted from P. K. Vinson, J. R. Bellare, H. T. Davis, W. G. Miller, and L. E. Scriven (1991). Direct imaging of surfactant micelles, vesicles, discs, and ripple phase structures by cryo-transmission electron microscopy. *J. Colloid Interface Sci.*, **142**, 75. Copyright 1991, with permission from Elsevier.

Figure 4.19    Reprinted from S. Rodriguez and H. Offen (1977). Micelle formation under pressure. *J. Phys. Chem.*, **81**, 47. Copyright 1977 American Chemical Society.

Figure 4.20    Replotted from G. D. Miles and L. Shedlovsky (1944). Minima in surface tension-concentration curves of solutions of sodium alcohol sulfates. *J. Phys. Chem.*, **48**, 57. Copyright 1944 American Chemical Society.

Figure 4.27    Adapted from G. T. Barnes (1978). Insoluble monolayers and the evaporation coefficient of water. *J. Colloid Interface Sci.*, **65**, 566. Copyright 1978, with permission from Elsevier.

Figure 5.10    Reprinted from B. F. Sinnamon, R. A. Dluhy, and G. T. Barnes (1999). Reflection-absorption FT-IR spectroscopy of pentadecanoic acid at the air–water interface. *Colloids and Surfaces A*, **146**, 49. Copyright 1999, with permission from Elsevier.

Figure 5.13    Reprinted with permission from G. T. Barnes, I. R. Gentle, C. H. L. Kennard, J. B. Peng, and I. McL. Jamie (1995). Interaction of phosphotungstate ions with phospholipid monolayers: a synchrotron X-ray study. *Langmuir*, **11**, 281. Copyright 1995 American Chemical Society.

Figure 5.15    N. K. Adam and G. Jessup (1926). The structure of thin films. Part VIII. *Proc. Roy. Soc. A*, **112**, 362. Replotted from Figure 1 with permission from The Royal Society (London).

Figure 5.18
Figure 5.20    Replotted from I. S. Costin and G. T. Barnes (1975). Two-component monolayers II. *J. Colloid Interface Sci.*, **51**, 106. Copyright 1975, with permission from Elsevier.

Figure 5.22    Replotted from M. A. McGregor and G. T. Barnes (1991). The equilibrium penetration of monolayers, V. *J. Colloid Interface Sci.*, **65**, 291. Copyright 1991, with permission from Elsevier.

Figure 5.24
Figure 5.25    Reprinted from J. B. Peng, G. T. Barnes, I. R. Gentle, and G. J. Foran (2000). Superstructures and correlated metal ion layers in Langmuir-Blodgett films of cadmium soaps observed with grazing incidence X-ray diffraction. *J. Phys. Chem.*, **104**, 5553. Copyright 2000 American Chemical Society.

Figure 5.26    Reprinted from J. B. Peng and G. T. Barnes (1994). The two-dimensional hexatic-B phase of single Langmuir-Blodgett monolayers observed by atomic force microscopy. *Thin Solid Films*, **252**, 44. Copyright 1994, with permission from Elsevier.

Figure 7.6    Reprinted from R. Wiesendanger, G. Tarrach, L. Scandella, and H. J. Guntherodt (1990). Scanning tunnelling microscopy. *Ultramicroscopy*, **32**, 291. Copyright 1990, with permission from Elsevier.

Figure 7.10    G. A. Somorjai (1994). *Introduction to Surface Chemistry and Catalysis*. Copyright John Wiley and Sons Limited. Reproduced with permission.

Figure 7.16    K. W. Kolasinski (2002). *Surface Science: Foundations of Catalysis and Nanoscience*. Copyright John Wiley and Sons Limited. Reproduced with permission.

Figure 9.2    J. J. Kipling and E. H. M. Wright (1962). *Journal of the Chemical Society (Resumed)* **1962**, 855–860, DOI: 10.1039/JR9620000855. Reproduced by permission of the Royal Society of Chemistry.

Figure 9.3    Replotted from R. S. Hansen and R. P. Craig (1954). The adsorption of aliphatic alcohols and acids from aqueous solutions by non-porous carbons. *J. Phys. Chem.*, **58**, 211. Copyright 1954 American Chemical Society.

Figure 9.4    Data from F. E. Bortell and Ying Fu (1929). Adsorption from aqueous solutions by silica. *J. Phys. Chem.*, **33**, 676. Copyright 1929 American Chemical Society.

Figure 9.14    Reprinted from J. N. Israelachvili, *Intermolecular and Surface Forces*, 2nd edn., Fig. 13.2. Copyright 1991, with permission from Elsevier.

Figure 9.16    Adapted from A. Ulman, *An Introduction to Ultrathin Organic Films from Langmuir–Blodgett to Self-assembly*, Fig. 3.5. Copyright 1991, with permission from Elsevier.

Figure 10.2    Replotted from D. Gorwyn and G. T. Barnes (1990). Interactions of large ions with phospholipid monolayers. *Langmuir*, **6**, 222. Copyright 1990 American Chemical Society.

Figure 10.9    Adapted with permission from S. J. Singer and G. L. Nicolson (1972). The fluid mosaic model of the structure of cell membranes. *Science*, **175**, 720. Copyright 1972 AAAS.

Table 2.2    Data from J. J. Jasper (1972). The surface tension of pure liquid compounds. *J. Phys. Chem. Ref. Data*, **1**, 841. Copyright 1972, with permission from the American Institute of Physics.

Table 4.1    R. C. Weast (1977). *Handbook of Chemistry and Physics, 58th ed.* Data reproduced with permission from CRC Press.

Table 6.1    Reprinted in part from L. A. Girifalco and R. J. Good (1957). A theory for the estimation of surface and interfacial energies. 1. *J. Phys. Chem.*, **61**, 904. Copyright 1957 American Chemical Society.

# Symbols and abbreviations

As in any field, there are certain accepted conventions used in symbols and abbreviations, and in this book we have tried to use accepted terminology where possible. Generally the SI system of units and recommended symbols has been used; however there are a few cases where additional symbols are needed or where some small modification to the recommendation is desirable. The symbols used are listed below:

## Symbols

| | | | |
|---|---|---|---|
| $A$ | total area of surface | $t$ | time |
| $\hat{A}_M$ | area per molecule of substance M | $U$ | internal energy |
| $a_i$ | activity of substance $i$ | $V$ | volume |
| $c_i$ | concentration of substance $i$ | $\overline{V}$ | partial molar volume |
| $D$ | diffusion coefficient | $V_a, V_r$ | potential energy of attraction, repulsion |
| $e$ | elementary charge | | |
| $E_k$ | kinetic energy | $v$ | rate |
| $E_b$ | binding energy | $w$ | work energy |
| $F$ | force | $x_i$ | mole fraction of substance $i$ |
| $G$ | Gibbs free energy | $x$ | distance in the surface plane |
| $g$ | acceleration due to gravity | $y$ | distance in the surface plane |
| $H$ | enthalpy | $z$ | distance normal to the surface plane |
| $I$ | ionic strength (Eq. 9.17) | $z_i$ | charge number of substance $i$ |
| $M$ | molar mass | $\Gamma_i$ | adsorbed amount of substance $i$ on unit area of surface |
| $m_i$ | mass of substance $i$ | | |
| $N_A$ | Avogadro constant | $\gamma$ | surface tension |
| $N_i$ | number of molecules of substance $i$ | $\zeta$ | zeta (electrophoretic) potential |
| $N_p$ | number of particles | $\eta$ | viscosity |
| $n_i$ | amount (in moles) of substance $i$ | $\mu_i$ | chemical potential of substance $i$ |
| $n_i^\sigma$ | total adsorbed amount (in moles) of substance $i$ | $\theta$ | fraction of surface covered by adsorbed film; contact angle |
| $P$ | pressure | $\Pi$ | surface pressure |
| $p$ | partial pressure or vapour pressure | $\Pi^{eq}$ | equilibrium spreading pressure |
| $q$ | heat energy | $\Pi^{os}$ | osmotic pressure |
| $R$ | gas constant | $\pi$ | ratio of circle circumference to diameter $= 3.1416$ (to four places) |
| $r$ | radius; transport resistance | | |
| $S$ | entropy; spreading coefficient | $\rho$ | density |
| $T$ | temperature | | |

## Superscripts and subscripts

| | |
|---|---|
| $\bullet$ | indicates a pure substance |
| $\alpha, \beta$ | indicate a bulk phase |
| S, L, G | indicate solid, liquid, or gas phase |
| $\sigma$ | indicates a surface excess quantity |
| $\theta$ | indicates a standard state |

## Abbreviations

| | |
|---|---|
| AFM | atomic force microscopy |
| BAM | Brewster angle microscopy |
| BET | Brunauer, Emmett, and Teller (equation for gas adsorption) |
| cmc | critical micelle concentration |
| $C_{16}TAB$ | cetyl trimethyl ammonium bromide |
| $C_nOH$ | alcohol with $n$ carbon atoms in chain |
| FTIR | Fourier transform infrared (spectroscopy) |
| GIXD | grazing incidence X-ray diffraction |
| LB | Langmuir–Blodgett (technique for depositing films on solids) |
| PEO | poly(ethylene oxide) |
| QCM | quartz crystal microbalance |

# 1 Introduction

## 1.1 The importance of interfaces

Interfaces are everywhere: in our bodies, in the food we eat and the drinks we drink, in plants, animals, fish, insects and microbes, in our cars, in the soil, in the atmosphere, in manufacturing and chemical factories. Occasionally the presence of an interface does not appreciably alter the behaviour of a system, but in many cases it does have a significant effect and in some cases it dominates the behaviour.

Think for instance of the catalytic converter, which has been an integral part of cars produced over the last decade. As the raw exhaust gases from the engine pass over the solid metal and metallic oxide catalyst surface, a large number of reactions take place, leading to more complete oxidation of the gases and hence to a cleaner environment. The gases spend only milliseconds in contact with the surface, and yet this has a dramatic effect on the composition of the final exhaust mixture that is vented to the atmosphere. As the contact between the gases and the surface is only really with the first couple of atomic layers, the properties of the solid support, other than ensuring that the surface area is very large, are largely irrelevant. It is the surface of the catalyst that is crucial.

Another important example concerns the inner lining of the lung. A layer of fluid lines the alveoli, and at the surface of this fluid, in contact with the air, is a layer that is only one molecule thick composed mainly of phospholipids. This mixture of phospholipids and some proteins is known as **lung surfactant**. Lung surfactant serves to lower the amount of work required for the action of breathing, a function that is so important that if the surfactant is not fully developed, unaided breathing is impossible. Because lung surfactant is only produced late in gestation, most infants who are born before 30 weeks gestation must be treated immediately after birth to ensure their survival. Again, it is the properties of the very outermost surface that determine the functioning of the entire system.

Again and again we encounter similar examples of the importance of surfaces. With the increasing emergence of new technologies relying on miniaturization, surface properties are growing in importance. The new and exploding field of nanotechnology is an obvious case where the solid surfaces of, for example, nanoparticles and mesoporous materials, are the places where the processes of interest take place. An understanding of processes occurring at surfaces is therefore relevant to many new developments.

Even though surfaces are increasingly important to modern technology, they have in fact been studied for a very long time. Reports dating from Roman times describe the calming of water by spreading oil on the surface, but arguably the beginning of the field as a scientific discipline dates from the experiments of Benjamin Franklin, reported to the Royal Society in 1774, where he describes the spreading of oil on a pond in Clapham Common, on the outskirts of London. He observed that placing as little as one teaspoon of oil on the surface calmed the ripples on a small area that quickly extended to about half an acre $(2 \times 10^3 \, \text{m}^2)$.

Before we can describe the complex processes that occur at surfaces, it is important to describe exactly what we mean by this term.

## 1.2 Surfaces and interfaces

### 1.2.1 Introduction

Where two homogeneous bulk phases meet there is a region of finite thickness where the properties change, often markedly, as we move from one bulk phase to the other. Such regions are known as surfaces or interfaces (although the term interphase would be a better description). Although we may commonly think of a surface as being of negligible thickness, in fact when we are discussing phenomena at a molecular level the thickness of the interfacial region is significant and definitely non-zero.

The properties of the interfacial region are particularly important when one of the phases is dispersed as many very small particles in the other phase, because of the dramatic increase in surface area. The two phases are usually referred to as the disperse phase and the continuous phase. Examples include colloids, emulsions, aerosols, and some natural and synthetic polymers. Often the particles are below the resolution limit of the optical microscope ($<0.5 \, \mu\text{m}$) but above the molecular size range. Table 1.1 illustrates how the surface area increases if we take a disperse phase of total volume $1 \, \text{cm}^3$ and subdivide it into smaller and smaller cubic particles.

Thus, for example, $1 \, \text{cm}^3$ of disperse phase divided into cubic particles with sides of $0.1 \, \mu\text{m}$ (within the size range of most colloids) has an interfacial area of $60 \, \text{m}^2$.

**Table 1.1.** Effect of subdivision into cubic particles on the surface area of a disperse phase with a total volume of $1\,cm^3$.

| Number of particles | Particle volume/m$^3$ | Length of cube edge/m | Total surface area/m$^2$ |
| --- | --- | --- | --- |
| 1 | $10^{-6}$ | $10^{-2}$ | 0.0006 |
| $10^3$ | $10^{-9}$ | $10^{-3}$ | 0.006 |
| $10^6$ | $10^{-12}$ | $10^{-4}$ | 0.06 |
| $10^9$ | $10^{-15}$ | $10^{-5}$ | 0.6 |
| $10^{12}$ | $10^{-18}$ | $10^{-6}$ | 6 |
| $10^{15}$ | $10^{-21}$ | $10^{-7}$ | 60 |

(a)

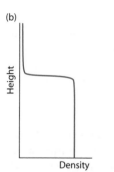

(b)

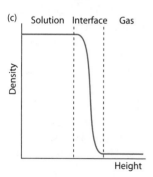

(c)

**Figure 1.1.** Density profile across the gas–liquid interface. The graph in (b) corresponds to the sample (a) and in (c) it has been rearranged into a more customary orientation. Dashed lines in (c) indicate the interfacial region.

## 1.2.2 Types of interface

Because there are three types of bulk phase, it is possible to classify interfaces based on the nature of the bulk phases that lie on either side of the interface. Thus there are five types of interface:

$$\left.\begin{array}{ll} \text{gas–liquid} & \text{G–L} \\ \text{liquid 1–liquid 2} & \text{L}_1\text{–L}_2 \end{array}\right\} \text{ fluid interfaces}$$

$$\left.\begin{array}{ll} \text{gas–solid} & \text{G–S} \\ \text{liquid–solid} & \text{L–S} \\ \text{solid 1–solid 2} & \text{S}_1\text{–S}_2 \end{array}\right\} \text{ non-fluid or solid interfaces}$$

Of course all gases mix with one another so there are no gas–gas interfaces. Usually with this notation the less dense phase will be shown first: G–L rather than L–G, for example.

When three bulk phases meet in a line, this line is known as a *triple interface*.

## 1.2.3 Defining the interfacial region

Taking an intensive property, such as density, and scanning its value from, say, a liquid phase through the interface to a gas phase would give a plot such as that in Figure 1.1.

In this case the density shows a smooth transition from the high density of the liquid to the much lower density of the gas. The bulk phases can be separated from the interface by two surfaces parallel to one another and positioned so that the bulk phases are homogeneous and uniform (uniform density in this case) while the inhomogeneity and non-uniformity are contained entirely within the interfacial region lying between the two surfaces. The dashed lines in Figure 1.1(c) illustrate this point.

We will see later that for some properties the transition from one bulk phase to another does not follow a smooth monotonic transition such as that in Figure 1.1. For example, the concentrations of some solutes (particularly marked with the surfactants, see Section 4.6) at the gas–solution interface

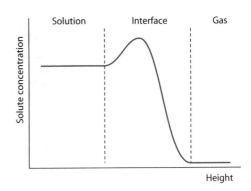

**Figure 1.2.** Concentration profile for a solute at the gas–solution interface.

may reach values very much higher than those in either bulk phase and exhibit a profile such as that in Figure 1.2.

The solute in Figure 1.2 is said to be *adsorbed* at the interface. This term will be defined and discussed in more detail in Chapter 3.

## 1.3 Stable interfaces

For any system at equilibrium, the free energy is at a minimum. If the system contains an interface, it is reasonable to expect that the interface would contribute to the free energy, and that this contribution would be a function of the area, $A$. We might expect that this contribution would take the form:

$$G = \gamma A + \text{other terms.} \tag{1.1}$$

In this equation, the coefficient $\gamma$ is known as the **surface tension** or **interfacial tension**. If the system is stable it follows that $\gamma$ must be positive, for if it were negative, an increase in area would lead to a lowering of the free energy, and therefore the surface area would spontaneously expand. This would ultimately lead to the dissolving of one substance in the other. Figure 1.3 illustrates how this process might occur for two liquids. In fact, the opposite occurs for immiscible liquids, suggesting that for a stable interface $\gamma$ must be positive and that at equilibrium the interfacial area will tend to a minimum in order to minimize the free energy. This process provides the driving force for many of the phenomena that will be discussed in following chapters.

Equation (1.1) also indicates that processes that lower the value of the interfacial tension would also be thermodynamically favoured. This is, then, a second force capable of driving interfacial processes. The phenomenon of adsorption is an important example.

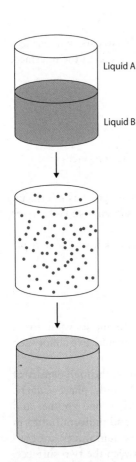

**Figure 1.3.** The spontaneous increase in surface area that would occur between two liquids if the surface tension were negative. This does not occur for immiscible liquids!

## 1.4 Key concepts

Although the topic of interfacial science is vast, we have attempted to introduce the crucial areas in the subsequent chapters. In this section we give a broad overview of the field.

### 1.4.1 Surface tension

Probably the most important concept of all, particularly when dealing with interfaces in which both phases are fluids, is that of surface tension. The existence of surface tension, and the effects which arise as a consequence, play a major role in the behaviour of systems containing interfaces. Because surface tension represents extra energy which is proportional to the area, systems attempt to minimize their surface area, resulting in the familiar fact that drops of liquid in air and bubbles are spherical. Many effects are more subtle, however, and are only evident if the interface is curved.

### 1.4.2 Wetting

The shapes of liquid droplets and the wetting of solid surfaces are determined principally by the forces acting at the relevant interfaces. These forces determine, for example, whether a liquid drop will spread over a solid surface or roll up into a ball, whether liquid will rise up the narrow gaps between the fibres of a wick, whether water will penetrate through the weave of an umbrella cloth, and contribute to the rise of sap in the stems of plants.

We are all familiar with the non-stick properties of Teflon$^{TM}$, which benefits from the tendency of liquids not to wet the surface, and fabrics designed to be non-wetting have the potential to be self-cleaning and never need washing!

### 1.4.3 Adsorption

A major consideration of interfaces is that the two phases normally have quite different properties, and frequently materials that are soluble in one or both phases will find it energetically favourable to concentrate (or deplete) at the interface. The main example is the class of materials called surfactants, which are molecules that are designed to have a part that prefers one phase and another part that prefers the other. Such materials adsorb at the interface, and this is crucial to a number of processes, such as detergency, in which oils can be dispersed in water when naturally they are very insoluble.

Adsorption is not restricted to liquid–liquid interfaces. Gases adsorb on to solids, a fact which is exploited in the routine measurement of surface areas of powders, and is an essential part of the catalysis of many gas phase reactions such as those that occur in the catalytic converters of nearly all cars. Such catalysis is also essential in many aspects of chemical industry.

Adsorption also leads to methods of fabrication of thin films, as thin as a single molecular layer. Methods exist which begin with a single layer at the air–water interface (the Langmuir–Blodgett technique) to build up multilayer films one layer at a time, and other methods use the ability of molecules to adsorb and self-assemble at the interface of a solution and a solid surface (known as self-assembled monolayers). Monolayers on an air–water interface can also be used to retard the evaporation of water, which is a major concern with open water storage dams in dry climates.

### 1.4.4 Emulsions

Emulsions, defined loosely as small droplets of one immiscible liquid in another, are very common, although frequently we might not recognize their presence. Common household products in which emulsions are frequently present are foods, paints, and cosmetics. To take paints as an example, these days most paints are water based. The water acts as a dispersion medium for an emulsion of polymer particles. After application, the water evaporates and the polymer particles coalesce to form a protective film.

Because emulsions are inherently unstable, preferring to minimize the surface area of the interface by individual droplets merging together, there is an enormous amount of science in the stabilization of emulsions in order to improve the effectiveness and marketability of products. This field is closely related to adsorption and surfactants, as the stabilization of emulsions relies on the use of surfactants in nearly all cases.

### 1.4.5 Colloids

Closely related to emulsions, colloids are very small solid particles in a liquid medium. They are also very common in industry and household products, and even particulates in air pollution represent a colloidal dispersion. Like emulsions, the major issue with colloids is their tendency to aggregate and achieving stability is usually the goal in formulation of products. This requires a detailed understanding of the charged nature of the solid–liquid interface, and the effects of added salts and other conditions.

### 1.4.6 Membranes

No biological organisms as we know them would exist without the existence of cell membranes, which separate two liquid phases which would otherwise mix. The other function of membranes is to allow the controlled movement of molecules into and out of cells, which is a remarkably complex task achieved by equally remarkable materials. These are primarily molecules known as lipids and special proteins called membrane proteins.

Knowledge gained from the study of biological membranes is increasingly being used to design better drug delivery systems.

## FURTHER READING

### The literature of interfacial science

There is an extensive literature dealing with surface chemistry and its various manifestations. This ranges from reports from Roman times of calming waves by the spreading of oil; through Benjamin Franklin's experiments (1774) with monolayers on a pond in London; the development of the first film balance by Agnes Pockels (1891) in the kitchen of her parents' home; the development of a theory to describe the adsorption of gases on solids by Brunauer, Emmett and Teller (1938); the formulation of a theory for the stability of lyophobic colloids by Deryaguin and Landau (1941) and Verwey and Overbeek (1948); cloud-seeding experiments by Langmuir and Schaefer (1946) and by Vonnegut (1946); and the experiments of Mansfield and Vines (1955 to 1962) on the retardation of water evaporation from large storages, to the many research and review papers currently being published in international journals of high repute.

A large number of books have been published on various aspects of interfacial science. Many of these will be found in the Further Reading sections near the end of each chapter of this book. Also listed under this heading are relevant review articles, mostly from one of the review journals or series devoted to our topic.

Original research is generally published in specialist journals, while there is also a small number of papers in more general journals. The titles of these journals can be seen in the references at the end of each chapter.

### Chapter references

In each chapter the references are shown by author and year of publication. For example, Peng *et al.* (2001). A book reference may also include a chapter or page number. At the end of the chapter the references to books and major review articles are listed alphabetically under Further Reading.

# 2 Capillarity and the mechanics of surfaces

Many commonly occurring phenomena that are observed in systems containing an interface in which one of the phases is a liquid can be understood through the concept of surface tension. Some examples are the rise of liquid in a narrow tube, the fact that drops of liquid tend to be spherical, and the observation that water spreads evenly on some surfaces while remaining in isolated drops on others. In this chapter, surface tension is defined and we see how it can be used to develop simple theories to explain these experiences.

## 2.1 Surface tension

The bristles in a wet paint brush tend to stick together, and we might be tempted to think that the presence of water is sufficient to make the bristles stick to one another. However, if the brush is held completely under water the bristles separate (Figure 2.1), so it is not the fact that the bristles are wet, but the presence of the air–water interface that causes them to stick together.

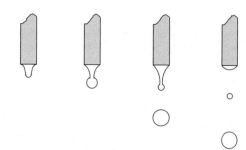

Dry

Wet

Immersed

**Figure 2.1.** Effects of water on the fibres of a brush.

**Figure 2.3.** The tension in the rubber acts on a line in the surface, causing it to stretch.

**Figure 2.2.** Spherical drops falling from a round tube.

Another example is that it is easy to make sandcastles with damp sand, but if the sand is dry or very wet it doesn't hold together. In both cases, the stickiness depends on the presence of the air–water interface, and the phenomena can be explained by realizing that the interface acts as though it were under tension. That is, it experiences forces that pull the bristles or sand particles together.

Another common example is that a drop of water tends to assume a spherical shape (distorted perhaps by gravity or by air resistance if falling, Figure 2.2). Now a spherical shape has the lowest surface area for a given volume of liquid, so what we observe in these and other examples is that the area of an interface tends to a minimum. The force that causes this to happen is called the *surface tension* or sometimes the *interfacial tension*.

If an imaginary line is drawn on a surface it will be pulled by the surfaces on either side towards those surfaces. For example, consider a partly inflated balloon (Figure 2.3). If a line is drawn on the surface, and then more air is added to the balloon, we observe that the line broadens as the rubber expands, and if the balloon were to be cut along the line the two sides would separate forming a hole.

The balloon is only an analogy, but something similar happens at the surfaces of interest to us. At any surface, if the force, $F$, acting tangentially to the surface and at right angles to an element, $\delta x$, of an imaginary line in the surface has a magnitude that is independent of the direction of the element, then the surface tension, $\gamma$, is:

$$\gamma = \frac{F}{\delta x}. \tag{2.1}$$

In words, the surface tension is the force per unit length acting on an imaginary line drawn in the surface. The SI units of surface tension are $N\,m^{-1}$, although because the $N\,m^{-1}$ is rather large (the surface tension of the air–water interface at room temperature is about $0.072\,N\,m^{-1}$), surface tension is more commonly quoted in $mN\,m^{-1}$.[1]

An intuitive way to understand the origin of surface tension from a molecular point of view is to consider the forces acting on a molecule at the surface of a liquid, compared to those acting on one in the bulk (Figure 2.4).

---

1. The c.g.s unit for surface tension is the $dyn\,cm^{-1}$, where $1\,dyn\,cm^{-1} = 1\,mN\,m^{-1}$. This unit is still occasionally seen in the literature.

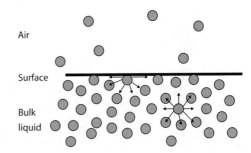

**Figure 2.4.** Forces acting on molecules near a surface.

The attractive forces acting on one molecule in the bulk are, when averaged over time, isotropic. That is, there is no net force pulling the molecule in any given direction. A molecule at the surface, however, will feel an unbalanced force due to the relative scarcity of near neighbours in the direction of the gas phase. The result is that there is a tendency for that molecule to be pulled into the bulk, as is the case for every other molecule at the surface. Hence it is clear why the tendency of the system is to minimize the area of the surface.

## 2.2 **Work of extension**

As the area of an interface tends to a minimum, energy must be brought into the system to extend the interface.

A surface contained by a rectangular frame with one end capable of sliding along the sides (Figure 2.5) exerts a force, $F$, on the slide. The length of the line of contact between soap film and slide is twice the length of the slide (i.e. $2(\frac{1}{2}x) = x$) because the soap film has two sides. That is, if you think of the film as being like this page, there are two interfaces, one on the reader's side of the page, and one on the reverse side of the page. If the surface is extended by moving the slide through a distance $\delta y$ then the area increases by $\delta A$ $(= x\,\delta y)$ and the force exerted by the surface tension and resisting the extension is (from Eq. 2.1):

$$F = \gamma x.$$

The work $(w_s)$ of extension is therefore:

$$w_s = F\,\delta y$$
$$= \gamma\,x\,\delta y$$
$$= \gamma\,\delta A. \tag{2.2}$$

This relationship is the basis for developing the thermodynamics of surfaces, a topic that will be explored in the next chapter.

In this argument it has been assumed that the new interface which is created when the surface area is increased has the same composition as the original. This requires new material to be brought from the bulk phases. If, instead, the original interface is stretched with no new material brought to the surface, additional work would be needed to increase the molecular separation.

If the surface of a solid is extended, then in general the work required is performed against the *interfacial stress*, rather than against the surface tension.

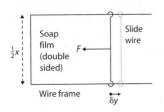

**Figure 2.5.** Wire frame for extending the surface of a (double-sided) soap film.

The interfacial tension is one component of the interfacial stress, but the relationship between these quantities is complex and depends on the nature of the solid with such factors as the isotropic or non-isotropic nature of the surface, the presence of defects, and so on. Methods for measuring the surface tensions of solids are at best approximate and often involve questionable assumptions. For example, the measured increase in vapour pressure for small spherical particles of solid can be used with the Kelvin equation (to be discussed later in this chapter, Eq. 2.19), but the assumptions of sphericity and uniformity of the particles are usually questionable. Further consideration of this topic is outside the scope of this book. More information can be obtained from Lyklema (2000) and Adamson and Gast (1997).

## 2.3 Contact angle, wetting, and spreading

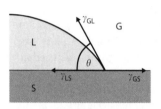

**Figure 2.6.** The forces acting at the triple interface for a drop of liquid on a flat solid surface.

When a drop of liquid is placed on a solid surface the triple interface formed between solid, liquid, and gas will move in response to the forces arising from the three interfacial tensions until an equilibrium position is established. The situation is illustrated in Figure 2.6 which shows a drop of liquid (L) on a flat solid surface (S) with air (G) as the third phase. The angle, $\theta$, between the solid surface and the tangent to the liquid surface at the line of contact with the solid is known as the **contact angle**. By convention the contact angle is measured in the liquid phase. The largest observed value for water on a smooth solid surface at equilibrium is about 120°.

The position of the triple interface will change in response to the horizontal components of the interfacial tensions acting on it. At equilibrium these tensions will be in balance and thus:

$$\gamma_{GS} = \gamma_{LS} + \gamma_{GL} \cos \theta. \tag{2.3}$$

However, when the drop is initially placed on the surface the interfacial tensions will not be in equilibrium. The net force per unit length of the triple interface along the solid surface will then be:

$$F_h = \gamma_{GS} - \gamma_{LS} - \gamma_{GL} \cos \theta' \tag{2.4}$$

where forces acting to the right are given positive signs and those acting to the left have negative signs. The angle $\theta'$ is the instantaneous contact angle and will change as the triple interface moves towards its equilibrium position where $\theta' = \theta$ and $F_h = 0$.

Although this equation describes the equilibrium contact angle in terms of the interfacial tensions involved, it gives no real insight into the *reason* that a certain value of contact angle is reached. We observe that some surfaces have a very high contact angle for water while for others it is so low as to be unmeasurable. An understanding of the origin of contact angle requires knowledge of the balance of forces between molecules in the liquid drop (*cohesive* forces) and those between the liquid molecules and the surface (*adhesive* forces). A surface that has primarily polar groups on the surface, such as hydroxyl groups, will have a good affinity for water and therefore strong adhesive forces and a low contact angle. Such a surface is called *hydrophilic*. If the surface is made up of non-polar groups, which is common for polymer surfaces or surfaces covered by an organic layer, we say that the surface is *hydrophobic*, and the contact angle will be large. For this reason, measurements of contact angle are frequently used as a quick and simple method to gain qualitative information about the chemical nature of the surface.

Many surfaces show an apparent hysteresis, where different values of the contact angle are measured depending on whether the measurement is performed on a drop of increasing size (the *advancing* contact angle) or of diminishing size (the *receding* contact angle). This is normally due to roughness of the surface, and the difference between advancing and receding contact angles is another useful and quick method of gaining qualitative information about the nature of the surface. Although the macroscopic observation of angles shows hysteresis, it is likely that the microscopic angles, if they could be observed, would show no hysteresis. The advancing contact angle (liquid moving over an apparently dry surface) is greater than the receding angle, thus when a contact angle close to zero is needed, as for example when surface tension is to be measured by a Wilhelmy plate (see Section 2.6.1), the surface of the solid is often roughened and a receding contact angle used.

Contact angle measurements are made by placing a drop of the liquid on a surface and viewing it with some type of magnifying lens (Figure 2.7). The angle can then be measured optically, usually by taking a digital image and using software to determine the angle.

The most difficult part of dealing with very high contact angles is keeping the drops in place long enough to be measured! The slightest tilt on the surface is enough to cause the drops to roll off.

Wetting is determined by the equilibrium contact angle, $\theta$. If $\theta < 90°$, the liquid is said to wet the solid; if $\theta = 0$, there is complete or perfect wetting; if $\theta > 90°$ ($\cos \theta < 0$), the liquid does not wet the solid.

The spreading of a liquid over a solid depends on the components of the interfacial tensions acting parallel to the solid surface at the line of contact, as in Eq. (2.4). The triple interface will move in the direction dictated by $F_h$ until a value of $\theta$ is reached at which $F_h = 0$, giving equilibrium, or, if such a

**Figure 2.7.** A contact angle image showing a drop of water on a PTFE surface. The needle used to deposit the drop is visible at the top.

### Self-cleaning surfaces

The contact angle of water on surfaces varies greatly depending on the chemical nature of the surface. Hydrocarbon surfaces (like the paint surface of a car after it's been polished with a wax polish) have high contact angles, as do fluoropolymers like Teflon™. Superhydrophobicity generally refers to surfaces with water contact angle $\theta > 150°$. Fabrication of superhydrophobic coatings has become a rapidly expanding area of research over the last few years, due mainly to the useful application of self-cleaning that such surfaces exhibit. The two key attributes to be considered in the design of such interfaces are the chemical composition and the micro- and nanostructure. Nanostructural roughness is manipulated to create superhydrophobicity from (chemically) low ($\theta < 90°$) contact angle materials. Typically the surface areas of such materials are 2.5 times greater than would be expected from an equivalent flat surface. This minimizes the use of commonly used but expensive fluorinated compounds which also achieve high contact angles by exploiting chemical interactions.

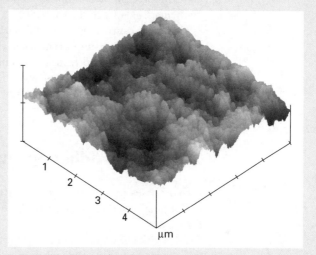

μm

The image above shows a three-dimensional view of a superhydrophobic surface viewed with an atomic force microscope. Although the contact angle of the underlying polydimethylsiloxane is about 75°, with the rough coating of nanoparticles the contact angle is 160–170° (courtesy of R. Lamb and H. Zhang).

position is not possible, the liquid spreads completely, and then $\theta = 0$, $\cos \theta = 1$, and

$$F_{h}(\theta = 0) = \gamma_{GS} - \gamma_{LS} - \gamma_{GL}. \tag{2.5}$$

It is convenient to define a spreading coefficient, $S_{LS}$, by

$$S_{LS} = \gamma_{GS} - \gamma_{LS} - \gamma_{GL}. \tag{2.6}$$

If $S_{LS} > 0$, the liquid spreads completely, whereas if $S_{LS} \leq 0$ the drop does not spread completely and it finds an equilibrium contact angle, $\theta^{eq}$, where $F_h = 0$.

## 2.4 The surface of tension

It has already been pointed out that any real interface has a finite thickness, but for many purposes it is convenient to replace the real interface by a mathematical surface of zero thickness which is mechanically equivalent to the real interface. This mathematical interface is called the surface of tension.

If it were possible to observe the pattern of forces acting on a plane cutting through an interface it would be complex and would extend over a region of finite thickness. Possibly a pattern such as that in Figure 2.8(a) might be found, but as the means for observing such a pattern do not exist some alternative is needed. In 1805 Young suggested that this complex pattern could be replaced by a mathematical surface under tension with a pattern of uniform forces in each of the bulk phases extending unchanged right up to the mathematical surface (Figure 2.8b).

This model surface is called the surface of tension. Proper location of the surface of tension is important when the interface is curved. In essence, two equations are obtained from the requirements that both the resultant forces and the resultant moments of those forces must be the same in the model system as in the real system. Details of the procedure have been described by Defay, *et al.* (1966).

## 2.5 Work of adhesion and cohesion

If two phases ($\alpha$ and $\beta$) in contact are pulled apart inside a third phase $\omega$, the original interface is destroyed and two new interfaces are formed (see Figure 2.9).

The work per unit area of each interface added to the system in performing this operation is called the work of adhesion, $w_{\alpha\beta}$:

$$w_{\alpha\beta} = -\gamma_{\alpha\beta} + \gamma_{\alpha\omega} + \gamma_{\beta\omega}. \tag{2.7}$$

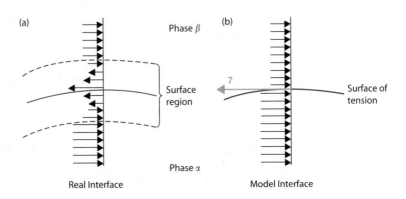

**Figure 2.8.** Possible pattern of forces in the real interface and the model interface.

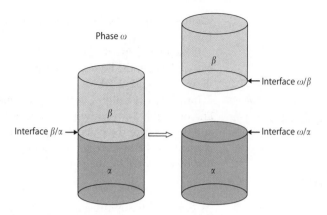

**Figure 2.9.** The separation of two phases.

If, instead of two distinct phases, a column of a single liquid is pulled apart, the work of cohesion is:

$$w_{\alpha\alpha} = 2\gamma_{\alpha\omega} \tag{2.8}$$

as $\gamma_{\alpha\alpha} \equiv 0$.

When one of the phases is a solid, the expression for work of adhesion (Eq. 2.7) can be combined with the equation for the contact angle (Eq. 2.3) to give

$$\begin{aligned} w_{LS} &= \gamma_{LG} + \gamma_{SG} - \gamma_{LS} \\ &= \gamma_{LG}(1 + \cos\theta). \end{aligned} \tag{2.9}$$

This is commonly known as the equation of Young (1805) and Dupré (1869). Its significance is that it relates the work of adhesion to the readily measured quantities, $\gamma_{LG}$ and $\theta$, rather than to the inaccessible interfacial tensions involving the solid surface.

## 2.6 Measurement of surface tension

Surface tension is a fundamental quantity in the investigation of fluid interfaces so its measurement is of great importance. Many methods exist for measuring the value of the surface tension of an interface, and the choice of method depends on the given system. The surface tension of the interface between a static, aqueous solution and air can be measured by a number of methods, but if, for example, the solution is viscous or the surface tension is changing rapidly, a careful choice of techniques must be made.

### 2.6.1 Wilhelmy plate

Possibly the easiest way to demonstrate the force arising from surface tension is to dip a flat plate through the surface of a liquid and measure the force acting on it. This is known as a Wilhelmy plate, and is named after the scientist who first used such a device to measure surface tension (Wilhelmy, 1863).

If the plate is perfectly wetted by the liquid a meniscus will form where it passes through the liquid surface giving a contact angle of 0°. The situation is shown in Figure 2.10.

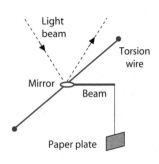

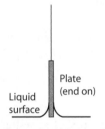

If the plate is hanging vertically, the meniscus will contact the plate along a line of length $2(x + y)$, where $x$ and $y$ are respectively the horizontal length and thickness of the plate. Along the line of contact the liquid surface will be vertical so the surface tension along this line will, from Eq. (2.1), exert a downward force on the plate of

$$F = \gamma \, 2(x + y). \qquad (2.10)$$

As $F$, $x$, and $y$ can all be measured, the surface tension can be determined.

Lane and Jordan (1970) have shown that despite the complex geometry of the meniscus at the edges of the plate, as shown in Figure 2.11, the simple formula of Eq. (2.10) holds. Thus the only correction that is needed arises from the buoyancy of the plate and that depends on the depth of immersion. If, however, the bottom edge of the plate is set level with the flat surface of the liquid the buoyancy correction is zero.

In the past, Wilhelmy plates have been made from a variety of materials, the most common being roughened mica, etched glass, and platinum. Scrupulous and elaborate procedures were needed with these materials to ensure that the surfaces were clean and that the contact angle was zero. In 1977 Gaines suggested the use of paper plates. With high quality filter or chromatography paper the liquid saturates the plate and essentially forms a liquid surface over the paper ensuring that the contact angle is zero.

Wilhelmy plates may be used in a static mode, as, for example, in a surface film balance (see Section 5.2.3), or in a detachment procedure where the difference in force between the plate hanging (wet) in air and at the moment of detachment when withdrawn from the liquid is measured. Detachment data require correction for buoyancy and there is usually some excess liquid clinging to the bottom edge of the plate. To some extent these two sources of error cancel, but nevertheless the Wilhelmy plate in detachment mode is only approximate.

The **du Noüy ring** is a variant of the Wilhelmy plate detachment technique. Instead of the flat plate a horizontal ring with a diameter of about 10 mm is

**Figure 2.10.** Wilhelmy plate for measuring surface tension. Here the force is measured by the twist of a simple torsion wire amplified by an optical beam. Electronic devices are normally used.

**Figure 2.11.** Photograph of the water meniscus on a Wilhelmy plate of roughened mica.

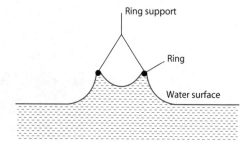

**Figure 2.12.** Cross-section of the surface for the du Noüy ring method for measuring surface tension.

used. However the assumption that the force attributable to surface tension is simply twice the circumference of the ring multiplied by the surface tension is an oversimplification because of the complex geometry of the surface (see Figure 2.12) so a correction procedure must be followed (Adamson and Gast, 1997).

## 2.6.2 Capillary rise

If a narrow capillary tube is dipped into a liquid the level of liquid in the tube is usually different from that in the larger vessel. With a clean glass capillary and a liquid that wets it, the liquid rises up the tube until an equilibrium position is attained (Figure 2.13).

At this point the liquid column in the capillary can be considered to be supported by the surface tension. For the situation where the liquid wets the capillary wall perfectly $(\theta = 0)$

$$\gamma 2\pi r_c = \Delta\rho g h \pi r_c^2$$
$$\gamma = \tfrac{1}{2}\Delta\rho g h r_c \qquad\qquad (2.11)$$

where $\Delta\rho$ is the difference in density between liquid and gas, $g$ is the acceleration due to gravity, $r_c$ is the radius of the capillary, and $h$ is the measured capillary rise. As these quantities are known or measurable the surface tension can be determined. The capillary rise is measured from the flat surface of the liquid in the large container to the height of the lowest part of the meniscus in the capillary (Figure 2.13). An alternative is to measure the difference in height between the rises of liquid in two capillaries of different internal diameters, which removes the difficulty of referring the height to a flat liquid surface.

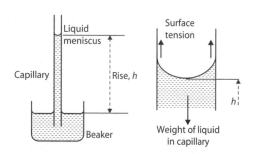

**Figure 2.13.** The phenomenon of capillary rise.

### The Thorny Devil

(*Moloch horridus*), a small lizard found in Australian deserts, uses a fascinating application of capillary rise to stay alive in extremely dry conditions. It is about 20 cm long, is slow moving, and lives mainly on ants. It is covered by grooved thorns, somewhat like rose thorns which form capillaries through which water is able to move over its body to the corners of its mouth from where it is able to drink. The water can come from the ground or from dew condensing on its body in the cold desert nights. Water is not absorbed through the skin.

The measurement of capillary rise and the use of Eq. (2.11) provide a simple and accurate means for determining the surface tension of a liquid. However it is essential that the capillary be scrupulously clean and that the contact angle be zero, as the accurate determination of $\theta$ in a capillary is extremely difficult. It is usually possible to meet these requirements with pure liquids, but with solutions there may be adsorption of the solute on the capillary walls leading to $\theta \neq 0$.

A more rigorous derivation is based on the Laplace equation (Section 2.7). It shows more clearly that for very precise measurements it is necessary to correct for the deviations of the liquid surface from a spherical shape. The procedures are too complex to be described here, but a detailed treatment can be found in Adamson and Gast (1997). An iterative procedure is used to refine the value of the capillary constant, $a$, which is defined by:

$$\frac{2\gamma}{\Delta \rho g} = a^2 = rh. \tag{2.12}$$

Here $r$ is the radius at the lowest part of the meniscus which is always slightly larger than the capillary radius, $r_c$. Actually the term *capillary constant* is misleading as its value depends on properties of the liquid, not the capillary.

### 2.6.3 Drop weight or volume

A drop of liquid hanging from the tip of a capillary is supported by the surface tension of the liquid. If we assume that the downward force due to the weight of the drop immediately before detachment is balanced by the upward force due to surface tension at the line of contact with the capillary tip we have:

$$mg = 2\pi r\gamma \tag{2.13}$$

where $m$ is the mass of the drop and $g$ is the acceleration due to gravity. The experimental procedure is to allow a known number of drops to form slowly and detach. The total weight of the detached drops is then measured. However there are serious errors in the assumption that the weight measured is the

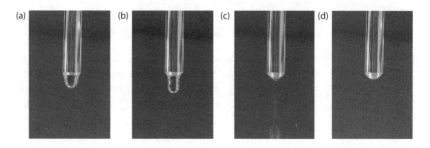

**Figure 2.14.** When a drop falls from a capillary, a significant amount of liquid remains (image d). This necessitates a correction to be applied to the simple formula of Eq. (2.13).

total weight of the drops, as a sizable fraction of the liquid hanging beneath the capillary remains attached to the capillary after each drop has fallen (Figure 2.14). Thus a correction factor, that depends on the capillary radius and the drop volume, has to be applied to Eq. (2.13).

Nevertheless the method is fast and easy to use. The drop volume method is a closely related one in which the drops are formed from the tip of a calibrated syringe. It is one of the few techniques that can be used at the liquid–liquid interface.

### 2.6.4 Maximum bubble pressure

During the formation of a bubble of an inert gas beneath a capillary dipping into a liquid the bubble radius is at first large, decreases to a minimum when the radius is the same as that of the capillary, and then increases again (Figure 2.15). Thus the pressure of gas in the bubble increases, passes through a maximum, and then decreases.

Because the pressure that the liquid exerts on the bubble varies with the depth of immersion the bubble shape is not accurately spherical and corrections similar to those used in capillary rise measurements are needed for accurate work.

Great care must also be exercised in the design and construction of the capillary tip. Usually a very sharp edge is used so that there is no ambiguity about the capillary radius on which the bubble is being formed.

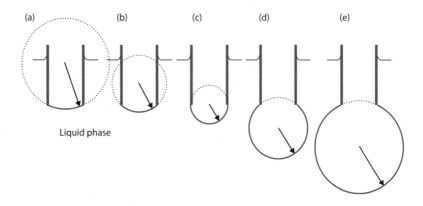

**Figure 2.15.** Shapes of a bubble in a liquid during development. Note that the radius of the bubble, indicated by the arrows, decreases to a minimum for case c, where the radius is equal to that of the capillary, and then increases again.

### 2.6.5 Sessile and pendant drops

The shape of a drop sitting on a flat surface (*sessile drop*) or hanging beneath a flat solid surface (*pendant drop*) is determined by the size of the drop and the surface tension of the liquid. Usually photographs of the drop are taken and measurements made on these. The methods are useful when only small volumes of the liquid are available.

### 2.6.6 Indicator oils

For measurements in situations where the use of laboratory instrumentation is not feasible, on reservoir surfaces for example, it is possible to estimate the surface tension by observing the spreading or otherwise of drops of various indicator oils. A drop of the oil is placed on the surface and its spreading behaviour observed. Usually this procedure is used to detect the presence of a surface film and to estimate the change in surface tension caused by the film.

For a drop of indicator oil placed on the surface of a liquid (the subphase), Eq. (2.6) shows that spreading will occur when

$$\gamma_{as} \geq \gamma_{ao} + \gamma_{os} \tag{2.14}$$

where a, o, and s refer to air, indicator oil, and subphase, respectively. For a given oil $\gamma_{ao}$ is fixed and known, and if we can assume that the oil drop displaces any surface film on the subphase $\gamma_{os}$ is also fixed and known. Consequently the equality in (2.14) gives the minimum value of the surface tension of the subphase, $\gamma_{as}$. Usually, however, it is the spreading coefficient, $S$, (Eq. 2.6) that is found in tables of data.

By using a set of indicator oils with different spreading coefficients (relative to a clean subphase surface) the surface tension of the subphase covered by a surface film can be estimated. Table 2.1 shows some potentially useful indicator oils for studying water surfaces. However it is important to note

**Table 2.1.** Indicator oils for estimating the surface tension of film-covered surfaces of water at 20 °C.

| Indicator oil | Minimum spreading tension, $\gamma_{min}$/mN m$^{-1}$ | Initial spreading coefficient, $S$/mN m$^{-1}$ |
|---|---|---|
| *n*-Octane | 72.6 | 0.15 |
| Benzene | 63.9 | 8.9 |
| *n*-Propyl palmitate | 57.2 | 15.5 |
| Aniline | 48.8 | 24.0 |
| Oleic acid | 48.1 | 24.6 |
| *n*-Octanol | 36.0 | 35.7 |
| *n*-Butanol | 26.2 | 46.5 |
| *iso*-Butanol | 22.8 | 49.9 |

that some of the liquids shown in Table 2.1 have a slight solubility in water and some can dissolve small amounts of water. Such solubility can alter the surface tension values and hence the spreading coefficient, so it is the initial spreading behaviour that should be observed, before appreciable dissolution has occurred.

Timblin *et al.* (1962) developed a set of indicator oils consisting of solutions of dodecanol in a light mineral oil. The spreading pressure is dependent on the dodecanol concentration and is approximately proportional to the logarithm of the concentration. Details of the mineral oil were not given so calibration would be needed before this procedure could be used with another oil.

## 2.7 The Laplace equation

If a fluid interface is curved the pressures on either side must be different. For example, if we take a flat soap film stretched across a circular frame the pressures on either side are the same (presumably atmospheric pressure). However in a soap bubble the pressure inside must be greater than that outside and we can demonstrate this by blowing the bubble on the end of a tube. It is necessary to blow into the tube to inflate the bubble and if the tube is left open the bubble will expel the air and shrink, eventually forming a flat film over the end of the tube.

When the system is at equilibrium, every part of the interface must be in mechanical equilibrium. For a curved interface, the forces of surface tension are exactly balanced by the difference in pressure on the two sides of the interface. This is expressed in the Laplace equation:

$$P^\alpha - P^\beta = \frac{2\gamma}{r} \qquad (2.15)$$

which holds for a spherical interface of radius $r$.

The derivation of the Laplace equation follows.

### Derivation of the Laplace equation

Consider a spherical cap, symmetrical about the $z$ axis and part of a spherical interface. The pressures exerted on the interface by the two bulk phases ($\alpha$ and $\beta$) will be different if the interface is curved and this difference ($\Delta P$) will give rise to a force acting along the normal to the interface at each point. The cap will also be subject to a force arising from surface tension acting tangentially at all points around the perimeter of the cap. These forces are shown in Figure 2.16(a).

#### Force arising from the pressure difference

For a small segment of cap of area $\delta A$ the force arising from the pressure difference is $(P^\alpha - P^\beta)\delta A$, and its component in the $z$ direction (the central axis of the cap) is $(P^\alpha - P^\beta)\delta A \cos\phi$.

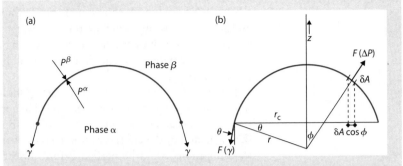

**Figure 2.16.** (a) Forces on a spherical cap; (b) resolution of forces.

But $\delta A \cos\phi$ is equal to the area of the projection of $\delta A$ on to the plane containing the perimeter of the cap. The sum of the resolved forces in the $z$ direction for the entire area of the cap is thus:

$$F_z^{\Delta P} = \left(P^\alpha - P^\beta\right)\sum \delta A = \left(P^\alpha - P^\beta\right)\pi r_c^2. \tag{2.15a}$$

**Force arising from surface tension**

The surface tension will exert a force, $\gamma\delta l$, on each element, $\delta l$, of the perimeter and acting tangentially to the surface at $\delta l$. The component in the $z$ direction is $\gamma\delta l \cos\theta$. But $\cos\theta = r_c/r$, so the component in the $z$ direction of the surface tension force on $\delta l$ is $\gamma\delta l \, r_c/r$. The sum over the whole perimeter is therefore:

$$\begin{aligned}
F_z^\gamma &= -\gamma \sum (\delta l)r_c/r \\
&= -\gamma(2\pi r_c)r_c/r \\
&= -2\pi r_c^2\gamma/r.
\end{aligned} \tag{2.15b}$$

The negative sign is used because this force acts in the opposite direction to the force arising from the pressure difference.

**Mechanical equilibrium**

If the system is in mechanical equilibrium the forces in the $z$ direction must sum to zero:

$$F_z^{\Delta P} + F_z^\gamma = 0$$

or

$$\left(P^\alpha - P^\beta\right)\pi r_c^2 - 2\pi r_c^2\gamma/r = 0 \tag{2.15c}$$

and so

$$P^\alpha - P^\beta = \frac{2\gamma}{r} \tag{2.15}$$

which is the Laplace equation for a spherical surface.

For a non-spherical interface two radii of curvature are needed. Figure 2.17 describes how the radii of curvature are defined.

By convention, positive values are assigned to the radii of curvature, $r$, $r'$, or $r''$, if they lie in phase $\alpha$.

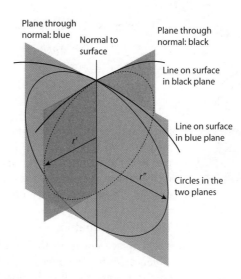

**Figure 2.17.** The two radii of curvature at a selected point ($r'$ and $r''$) are obtained by taking the normal to the surface at that point, drawing a plane containing the normal (in an arbitrary orientation) taking a second plane containing the normal at right angles to the first plane, observing the curves formed by the intersections of these planes with the surface and then finding the radii of the two circles drawn in these planes that have the same curvatures as these lines at the selected point.

The Laplace equation then becomes

$$P^\alpha - P^\beta = \gamma \left( \frac{1}{r'} + \frac{1}{r''} \right) = \frac{2\gamma}{r_m} \tag{2.16}$$

where $1/r_m$ is the mean curvature (inverse of the radius):

$$\frac{1}{r_m} = \frac{1}{2} \left( \frac{1}{r'} + \frac{1}{r''} \right). \tag{2.17}$$

### 2.7.1 Applications of the Laplace equation

#### The shapes of soap films

A soap film stretched across a flat wire frame is flat because the pressure is the same on both sides, but more complex shapes are possible even when there is no difference in pressure. For example, if the film is formed as an open cylinder (as in Figure 2.18) it will form into a 'wicker basket' shape where $r' = -r''$.

An interesting application of the Laplace equation is seen when two bubbles of different size are connected by a tube, as in Figure 2.19. It is often surprising that the small bubble shrinks while the larger bubble grows, but is perfectly understandable in the light of the Laplace equation, which predicts that the pressure inside the smaller bubble is greater than that in the larger one.

This and other experiments with soap bubbles have been described by Boys (1911).

#### Capillary rise

The *capillary rise* phenomenon can also be treated using the Laplace equation. Balancing the pressure decrease under the curved liquid surface (assumed to be spherical and making a contact angle of $\theta$ with the capillary wall) in the capillary

$$P^{atm} - P = \frac{2\gamma \cos \theta}{r_c}$$

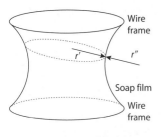

**Figure 2.18.** Sketch of a soap film stretched between two circular wire frames. If the top and bottom ends are open, the air pressure will be the same on both sides of the film so Eq. (2.15) requires that at any point on the surface $r'$ and $r''$ must be equal but with opposite signs.

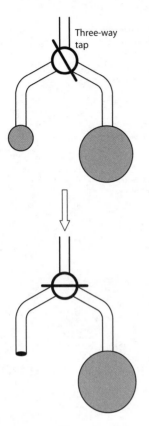

**Figure 2.19.** Illustration of the Laplace equation. When the tap is turned to connect the two bubbles (as in the lower figure) the smaller bubble shrinks and the larger bubble grows until the curvatures of both bubbles are the same. That is, the film over the end of the left tube has the same curvature as the large bubble.

with the pressure decrease from the weight of the liquid column:

$$P^{atm} - P = \Delta\rho g h$$

gives:

$$\gamma = \frac{\Delta\rho g h r_c}{2\cos\theta} \tag{2.18}$$

for a capillary of radius $r_c$ and a liquid with contact angle $\theta$. It is difficult to measure and to reproduce $\theta$ if $\theta \neq 0$ so only systems with $\theta = 0$ are used in practice. Deviations from the spherical shape are a complication for accurate measurements of surface tension and several correction procedures have been devised. These are described in detail by Adamson and Gast (1997).

### The Kelvin equation

The Laplace equation forms the theoretical basis for the Kelvin equation which describes the effect of surface curvature on vapour pressure. It is discussed in some detail in Section 2.8.

## 2.8 The Kelvin equation

An important consequence of the Laplace equation concerns the effect of surface curvature on the vapour pressure of a liquid. This relationship is known as the Kelvin equation:

$$\ln\left(\frac{p^c}{p^\infty}\right) = \left(\frac{\gamma \overline{V}^L}{RT}\right)\left(\frac{2}{r_m}\right) \tag{2.19}$$

where $p^c$ and $p^\infty$ are respectively the vapour pressures over the curved surface of radius $r_m$ and a flat surface ($r = \infty$). As can be seen from the derivation below, there is a convention of signs for $r_m$ that assigns a positive sign to the radius when it lies in the liquid phase and a negative sign when it lies in the vapour phase.

---

### Derivation of the Kelvin equation

At a curved interface the condition for mechanical equilibrium is given by the Laplace equation (2.15):

$$p^\alpha - p^\beta = \frac{2\gamma}{r_m} \tag{2.19a}$$

where $\alpha$ is the phase on the concave side. Generalize by giving $r$ a positive sign if it lies within phase $\alpha$, a negative sign if it lies within phase $\beta$.

For physico-chemical equilibrium:

$$\mu_i^\alpha = \mu_i^\beta = \mu_i. \tag{2.19b}$$

If the equilibrium is shifted slightly with adjustments to $p^\alpha$, $p^\beta$, $r_m$, and $\mu_i$, we obtain:

$$\delta p^\alpha - \delta p^\beta = 2\gamma\delta(1/r_m) \tag{2.19c}$$

and

$$\delta\mu_i^\alpha = \alpha\mu_i^\beta = \delta\mu_i. \tag{2.19d}$$

Since

$$\left(\frac{\partial\mu}{\partial p}\right)_{T,n_i,A} = \overline{V} \tag{2.19e}$$

we have:

$$\left(\frac{\partial \mu^\alpha}{\partial p^\alpha}\right)_{T,n_i^\alpha} = \overline{V}^\alpha, \quad \left(\frac{\partial \mu^\beta}{\partial p^\beta}\right)_{T,n_i^\beta} = \overline{V}^\beta. \tag{2.19f}$$

If the process is carried out at constant temperature and without transfer of material across the interface then (2.19d) and (2.19f) give:

$$\delta \mu_i = \overline{V}^\alpha \delta p^\alpha = \overline{V}^\beta \delta p^\beta. \tag{2.19g}$$

Substitution of (2.19g) into (2.19c) yields:

$$2\gamma\delta(1/r_m) = \delta p^\beta (\overline{V}^\beta - \overline{V}^\alpha)/\overline{V}^\alpha. \tag{2.19h}$$

If $\beta$ is the vapour phase,

$$\overline{V}^\beta \gg \overline{V}^\alpha$$

and (2.19h) becomes:

$$2\gamma\delta(1/r_m) = \delta p^\beta (\overline{V}^\beta/\overline{V}^\alpha) = \delta \mu_i/\overline{V}^\alpha \tag{2.19i}$$

where $\delta \mu_i$ refers to the process where the radius is changed. Integration of (2.19i) from a flat surface $(1/r_m = 0, \mu_i = \mu_i^\infty)$ to any selected curvature $(1/r_m, \mu_i^c)$ yields:

$$2\gamma(1/r_m) = (\mu_i^c - \mu_i^\infty)/\overline{V}^L$$

or

$$\mu_i^c - \mu_i^\infty = \gamma\overline{V}^L(2/r_m) \tag{2.19j}$$

where superscript $\alpha$ has been replaced by L to indicate the liquid phase. Equation (2.19j) is the general form of the Kelvin equation.

Note the approximation between (2.19h) and (2.19i) and the assumption that the liquid is incompressible when integrating (2.19i).

If the vapour behaves as an ideal gas,

$$\mu_i = \mu_i^\theta + RT\ln(p/p^\theta)$$

and

$$\mu_i^c - \mu_i^\infty = RT\ln(p^c/p^\theta). \tag{2.19k}$$

Equation (2.19j) now becomes:

$$\ln\left(\frac{p^c}{p^\infty}\right) = \left(\frac{\gamma\overline{V}^L}{RT}\right)\left(\frac{2}{r_m}\right) \tag{2.19}$$

which is the more usual form of the Kelvin equation.

### 2.8.1 Consequences of the Kelvin equation

The Kelvin equation has many important consequences as it provides explanations for such phenomena as the difficulties found in self-nucleation of a new phase, the growth of large droplets at the expense of smaller ones, and condensation in capillaries.

#### Droplet growth

For a spherical droplet in contact with its vapour, the two principal radii are equal, lie within the liquid phase and thus have positive signs. Hence, according to the Kelvin Eq. (2.19) the vapour pressure of the droplet will be greater than that of the same liquid with a flat surface; there is, for example, approximately a 10% increase in vapour pressure for a water droplet of radius 10 nm. The smaller the radius, the higher the vapour pressure so that if there are droplets of various sizes present the smaller ones will tend to evaporate while the larger ones will tend to grow. An important example occurs in clouds where the larger droplets grow until they are heavy enough to fall as rain.

A similar mechanism is thought to exist for crystals in a solution: the larger crystals tend to grow at the expense of the smaller ones leading to the process of digestion or **Ostwald ripening** which is used to increase efficiency in analytical and industrial filtration.

It is worth noting that the equilibrium between a small liquid droplet and its vapour is an unstable one. If the drop, originally at equilibrium, loses a few molecules and decreases in size its vapour pressure will increase above the actual vapour pressure, and the droplet will continue to evaporate. If, on the other hand, a few more molecules collide with the droplet increasing in size its vapour pressure will drop below the actual vapour pressure so that condensation will continue and the droplet will grow.

### Self-nucleation of a new phase

In self-nucleation or homogeneous nucleation of a new phase, the first step is the formation of very small nuclei or embryos of the new phase inside the old phase. Examples include the formation of minute liquid clusters in a supersaturated vapour and the formation of minute bubbles of vapour in a superheated liquid. However, if the experimental conditions are only marginally in favour of the new phase (e.g. if the supersaturation or superheating is only slight) the Kelvin equation predicts that the embryos will be unstable and will tend to disappear again. Only if the embryo is able to reach a certain critical size, given by substituting the actual vapour pressure in the Kelvin equation, will it be stable. Once the embryo exceeds this critical size it will grow rapidly.

### Heterogeneous nucleation

In most practical situations, solid surfaces, such as the vessel walls, dust particles, boiling chips, and so on, provide sites where heterogeneous nucleation of a new phase can occur under less extreme conditions than those required for homogeneous nucleation.

### Capillary condensation

Capillary condensation occurs when the adsorption of a vapour in a capillary forms a liquid surface with a very small radius of curvature. The radii then lie in the vapour phase and consequently the vapour pressure of the liquid is lower than that of the same liquid with a flat surface. Condensation rather than adsorption will occur if the actual vapour pressure is higher than the vapour pressure calculated from the Kelvin equation for the curved surface. This actual vapour pressure may well be lower than the saturation vapour pressure for a flat surface. This aspect will be discussed in more detail in Section 8.6.

## 2.8.2 Verification of the Kelvin equation

In order to verify the Kelvin equation experimentally it is necessary to measure both the vapour pressure and the surface curvature of a liquid with a highly curved surface. Convincing data have been obtained by La Mer and Gruen using monodisperse aerosols. Details of this procedure are given in Section 4.11.5.

Another approach uses the surface forces apparatus described in more detail in Section 9.9.6. In this apparatus the forces of interaction between two crossed cylinders of molecularly smooth mica can be measured as a function of separation. If the cylinders in contact with one another are exposed to the vapour of a test liquid, capillary condensation may occur at the junction if the vapour pressure is high enough. The relative vapour pressure can be controlled by setting the temperature of the bulk liquid or adding a non-volatile solute. With an annulus of condensed liquid at the junction the cylinders are slowly separated so that a bridge of liquid formed between the mica surfaces. When this bridge suddenly evaporates the separation is measured and from this $r_m$ can be calculated. It was found that the Kelvin equation held down to values of $r_m$ as low as 2.5 nm for several organic liquids. This result also indicates that macroscopic values of $\gamma$ and $\rho$ hold at such low dimensions. For water, however, mean radii lower than 5 nm do not follow predictions based on the macroscopic properties.

## 2.9 The surface tension of pure liquids

### 2.9.1 Effect of temperature on surface tension

The chief factor affecting the surface tension of a pure liquid is the temperature (for example, Figure 2.20). Experimentally it is found that the surface tension of a pure liquid decreases nearly linearly with temperature. It must, of course, be zero at the critical temperature, where the interface disappears.

For some liquids the empirical Eötvös equation describes the dependence of surface tension on temperature:

$$\frac{d\left(\gamma(M/\rho)^{2/3}\right)}{dT} = -2.12 \times 10^{-7} \text{J mol}^{-2/3} \text{K}^{-1} \tag{2.20}$$

where $M$ is the molar mass and $\rho$ is the density.

### 2.9.2 The surface tension of liquids

Values for the surface tensions of some pure liquids are given in Table 2.2.

The very high surface tension of mercury, typical of values for metals, is attributable to short-range electron exchange interactions, the origin of so-called *metallic bonds*.

It is notable that the surface tension of water is significantly higher than the surface tensions of all of the organic liquids. Only glycerol has a surface

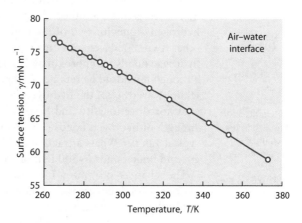

**Figure 2.20.** The surface tension of the air–water interface as a function of temperature. (Data from Vargaftik *et al.* 1983.)

**Table 2.2.** The surface tensions of some common pure liquids at the vapour–liquid interface at 20 °C. (From Jasper, 1972.)

| Liquid | Surface tension, $\gamma/\mathrm{mN\,m^{-1}}$ | Liquid | Surface tension, $\gamma/\mathrm{mN\,m^{-1}}$ |
|---|---|---|---|
| Acetic acid | 27.6 | n-hexane | 18.4 |
| Acetone | 25.1 | Isobutyl alcohol | 22.9 |
| Benzene | 28.9 | Methanol | 22.5 |
| n-butyl alcohol | 25.4 | Mercury | 486.5 |
| Carbon tetrachloride | 27.0 | n-octane | 21.6 |
| Chloroform | 27.2 | Oleic acid | 32.5 |
| Cyclohexane | 25.2 | Propanoic acid | 26.7 |
| Ethyl acetate | 24.0 | n-propyl alcohol | 23.7 |
| Ethanol | 22.4 | Pyridine | 37.2 |
| Di-ethyl ether | 17.0 | Toluene | 28.5 |
| Glycerol | 63.4 | Vinyl acetate | 24.0 |
| Ethylene glycol | 48.4 | Water | 72.8 |

tension approaching that of water. Water and glycerol (plus some other liquids not shown in Table 2.2) are highly polar hydrogen-bonding liquids and it is this feature that explains the high surface tensions. As water is a major component in most aspects of interface science, we will discuss it in detail below.

### 2.9.3 The hydrophobic–hydrophilic interaction

The structure of the water molecule is well known but it is important to consider here the distribution of electric charge in the molecule as this is the key to understanding its interactions with other molecules. The basic configuration is shown in Figure 2.21.

Water molecules tend to form hydrogen bonds with four other water molecules, giving a tetrahedral arrangement with each molecule providing the hydrogen atoms for two of the bonds and with the hydrogens atoms in the other two bonds coming from neighbouring water molecules. All four hydrogen bonds are approximately linear. In ice, this structure dominates, but although it tends to be retained in liquid water it is somewhat disordered and labile. Nevertheless the hydrogen bonds in liquid water have a strong tendency for directionality and linearity. It is also important to note that the strength of hydrogen bonds (10–40 kJ mol$^{-1}$) is appreciable greater than for typical van der Waals attraction ($\sim$1 kJ mol$^{-1}$) but much less than for covalent and ionic bonds ($\sim$500 kJ mol$^{-1}$) (Israelachvili, 1991, p. 127).

The charges on water molecules and their strong tendency to form hydrogen bonds govern their interactions not only with other water molecules but with other materials as well.

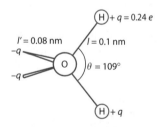

**Figure 2.21.** Sketch of water molecule showing the tetrahedral distribution of charge with bond lengths and angles. The charges ($q$) on the left represent the charges attributable to the unshared electron pairs. $e$ is the electronic charge. Based on the model of Stillinger and Rahman (1974).

If the other material is an ion, a polar compound, or a charged surface the neighbouring water molecules are able to arrange themselves with their charges of opposite sign towards the other material. This is the process of hydration. Energetically this is a very favourable arrangement. Solid surfaces at which there is extensive hydration are described as hydrophilic. In addition to these charged or polar entities, there are certain atoms that readily form hydrogen bonds when in appropriate compounds. Oxygen in alcohols and polyethylene oxide chains, and nitrogen in amines are important examples. These can interact with water by forming hydrogen bonds.

If, conversely, the other material is non-polar and incapable of forming hydrogen bonds the water molecules arrange themselves to minimize the number of unused potential hydrogen bonds. When the other molecule is small, the water molecules are able to arrange themselves into a cage-like structure around that molecule and so utilize all four potential hydrogen bonds per water molecule, forming a *clathrate* compound. This results in a decrease in entropy, as the water molecules are more ordered than in the bulk liquid, an effect that restricts the solubility of the solute. Important examples for the present context are the hydrocarbons and some long-chain hydrocarbon derivatives. For the hydrocarbons the entropy decrease is roughly proportional to the area of surface exposed to the water. Such substances are described as hydrophobic.

When two hydrophobic molecules encounter one another in an aqueous solution there would obviously be a decreased area of exposed hydrophobic surface if the two molecules came together. There is, therefore, an entropically driven tendency for such molecules to aggregate: a process known as hydrophobic interaction. The description of this attraction as a *hydrophobic bond* is not appropriate as no actual bond is formed and the attraction may occur over longer distances than for typical bonds.

At a solid or liquid surface formed from hydrophobic materials the water molecules must organize themselves to minimize the number of unused hydrogen bonds. Thus a drop of water on a hydrophobic surface tends to roll into a ball with a high contact angle (see Section 2.3) thereby minimizing contact with the surface and maximizing the formation of hydrogen bonds with other water molecules. The high surface tension of water can also be attributed to this effect as the air essentially forms a hydrophobic surface.

When a hydrophilic group and a hydrophobic group are combined in the one molecule we have a class of compounds now commonly referred to as amphiphiles (from Greek, meaning loves both, referring to water and to oil). In these molecules the presence of the hydrophilic group modifies the hydrophobic nature of the attached hydrophobic group. In long-chain alcohols, for example, the hydrophilic OH group significantly disrupts the tendency for a clathrate structure to form around the alkyl chain (Israelachvili, 1991, p. 135). Thus while insoluble monolayers can be formed with tetradecanol and longer chain alcohols, indicating negligibly low solubility, the shorter alcohols, methanol, ethanol, and propanol are fully miscible with water, with the solubility then decreasing rapidly with chain length.

## SUMMARY

The concept of surface or interfacial tension has been introduced to explain the observation that surfaces and interfaces tend to contract to the minimum area. Surface tension is defined as the force per unit length acting on an imaginary line drawn in the surface in a direction away from the line and tangential to the surface (Eq. 2.1). Surface tension can also be equated to the work of extending the surface by unit area, a definition that will lead to the thermodynamics of surfaces as discussed in Chapter 3.

Such tensions control the wetting of a solid surface by a drop of liquid, the adhesion between two phases, the pull of a liquid surface on a wetted solid dipping into it, the rise of liquid up a capillary tube, and other phenomena. Methods for measuring surface and interfacial tension are described.

When a surface is curved there is a difference in pressure on the two sides of the surface that is described by the Laplace equation (2.15). The Laplace equation is used to describe such phenomena as the shapes of soap films and capillary rise. It also leads to the Kelvin equation (2.19) which describes how the curvature of a liquid surface changes the equilibrium vapour pressure of the liquid. The consequences of the Kelvin equation are profound: it governs such processes as the growth of droplets and crystals, the self-nucleation of a new phase, and capillary condensation.

Finally the surface tension of pure liquids is discussed. Water has a significantly higher surface tension than most organic liquids and this is attributable to the formation of hydrogen bonds. Furthermore, it is the possibilities or otherwise for hydrogen bonding that determine the nature and strength of the interactions of liquid water with solutes and with surfaces.

## FURTHER READING

Adamson, A. W. and Gast, A. P. (1997). *Physical Chemistry of Surfaces, 6th edn.* Wiley, New York.

Boys, C. V. (1911). *Soap Bubbles: Their Colours and the Forces Which Mould Them.* Republished by Dover Publications, New York, 1959. Describes many simple experiments with soap films designed to generate interest and at the same time impart some science.

Defay, R., Prigogine, I., Bellemans, A., and Everett, D. H. (1966). *Surface Tension and Adsorption.* Longmans, London.

Israelachvili, J. N. (1991). *Intermolecular and Surface Forces, with Applications to Colloidal and Biological Systems, 2nd edn.,* Academic Press, London. Provides a somewhat different perspective on surface phenomena.

Lewis, G. N., Randall, M., Pitzer, K. S., and Brewer, L. (1961). *Thermodynamics, 2nd edn.* McGraw Hill, New York.

Lyklema, J. (2000). *Fundamentals of Interface and Colloid Science, III, Liquid–Fluid Interfaces.* Academic Press, San Diego.

## REFERENCES

Gaines, G. L. (1977). *J. Colloid Interface Sci.*, **62**, 191.

Jasper, J. J. (1972). *J. Phys. Chem. Ref. Data*, **1**, 841.

Lamb, R. N., Zhang, H., and Raston, C. L. (1997). Hydrophobic coatings. WO 9842452. Zhang, H., Jones, A. W., and Lamb, R. N. (2000) Hydrophobic coating material containing modified gels. WO 2001014497.

Lane, J. and Jordan, D. O. (1970). *Aust. J. Chem.*, **17**, 7.

Stillinger, F. H. and Rahman A. (1974). *J. Chem. Phys.*, **60**, 1545.

Timblin, L. O., Florey, Q. L., and Garstka, W. U. (1962). In V.K. La Mer, ed., *Retardation of Evaporation by Monolayers: Transport Processes*. Academic Press, New York, p. 177.

Vargaftik, N. B., Volkov, B. N., and Voljak, L. D. (1983). *J. Phys. Chem. Ref. Data*, **12**, 817.

Wilhelmy, L. (1863). *Ann. Physik.*, **119**, 177.

## EXERCISES

**2.1.** Calculate the work required to stretch a soap film so that its area increases by $4.0 \, cm^2$. The surface tension of the air–soap solution interface is $28 \, mN \, m^{-1}$.

**2.2.** Describe the spreading behaviour you would expect when a drop of benzene is placed on a clean water surface. For the pure liquids the interfacial tensions (in $mN \, m^{-1}$) are:

| | |
|---|---|
| air–water | 72.8 |
| air–benzene | 28.9 |
| benzene–water | 35.0 |

However small amounts of each liquid slowly dissolve in the other and this affects the interfacial tensions which then become:

| | |
|---|---|
| air–water | 62.2 |
| air–benzene | 28.8 |
| benzene–water | 35.0 |

**2.3.** A spherical drop of oil of radius $1.0 \, cm$ is dispersed as an emulsion in water. The oil particles in the final emulsion are spherical and have a radius of $0.1 \, \mu m$. Calculate the work required for the hypothetically reversible dispersion process assuming that the interfacial tension remains constant at $30.0 \, mN \, m^{-1}$ throughout.

**2.4.** A syringe needle (circular in section) with a sharp tip of radius $r$ is used to form air bubbles beneath the surface of a solution of surface tension $\gamma$. Derive an expression for the maximum pressure, $p$, observed during the slow formation of an air bubble at the tip of the needle. Assume that the bubble is a partial sphere. Show by a qualitative argument that the formula does indeed correspond to the maximum pressure.

**2.5.** Use the Kelvin equation to calculate the radius of an open ended tube within which capillary condensation of nitrogen at $77 \, K$ and a relative pressure of 0.75 might be expected. Assume that prior adsorption of nitrogen has formed a layer $0.9 \, nm$ thick coating the inside of the tube. For nitrogen at 77 K the surface tension is 8.85 mN m$^{-1}$ and the molar volume is 34.7 cm$^3$ mol$^{-1}$.

**2.6.** A jet aircraft is flying through a region where the air is 10% supersaturated with water vapour (i.e. the relative humidity is 110%). After cooling, the solid smoke particles emitted by the jet engines adsorb water vapour and can then be considered as minute spherical droplets. What is the minimum radius of these droplets if condensation is to occur on them and a 'vapour trail' form? Data: $\gamma(H_2O) = 75.2$ mN m$^{-1}$; $M(H_2O) = 0.018$ kg mol$^{-1}$; $\rho(H_2O) = 1030$ kg m$^{-3}$; $T = 275$ K.

# 3 Adsorption and the thermodynamics of surfaces

## 3.1 Introduction

The term adsorption can be loosely interpreted as the tendency for one component of a system to have a higher (or lower) concentration at the interface than it has in either of the adjacent bulk phases. Sometimes this tendency can be observed directly, for example as a decrease in pressure when a gas is adsorbed on to a solid, or a decrease in solute concentration when a solution is shaken with a solid powder. In other cases adsorption must be inferred from other information, such as a change in the surface tension of the solution.

However, before this concept of adsorption can be refined so that a quantitative measure can be developed, we need to consider the ways in which the interface is to be described.

## 3.2 Models of the interface

A system containing an interface can be divided into three regions: bulk phase $\alpha$, bulk phase $\beta$, and the interface $\sigma$. Thus the total value of an extensive property $B$ (e.g. $G, U, S, n_i, \dots$), has three components:

$$B = B^\alpha + B^\beta + B^\sigma. \tag{3.1}$$

The total value, $B$, is often known, but the subdivision of the system into the three contributions is, to some extent, arbitrary. Two approaches are in common use: the surface phase approach, and the surface excess approach. These are illustrated in Figure 3.1, where the intensity of $B, b$ ($= B/V$) is plotted against distance from the interface.

### 3.2.1 The surface phase approach

In this approach, a distinct surface phase is defined by two mathematical boundaries, one on either side of and parallel to the interface. These boundaries must be situated sufficiently far out from the interface for all of the perturbations of properties associated with the interface to be contained within the interfacial region. The regions assigned to the bulk phases are therefore uniform in their properties, that is, the intensive properties are constant throughout. Furthermore the interfacial region has an appreciable thickness. This approach is illustrated in Figure 3.1(b).

### 3.2.2 The surface excess properties approach

This approach is an extension of Young's model of the interface which introduced the surface of tension (Section 2.4). The interface is treated as a mathematical surface of zero thickness and the intensive properties of the bulk phases are assumed to retain their bulk values right up to this surface. This pattern constitutes the 'model' interface and invariably differs from the pattern at the 'real' interface. The difference between the values of the property for the entire system using the real ($B_{real}$) and the model ($B_{model}$) interfaces is termed the surface excess of the property and is assigned to the interface. Thus

$$B_{excess} = B^\sigma = B_{real} - B_{model}. \tag{3.2}$$

Note that all of the volume is assigned to the bulk phases:

$$V = V^\alpha + V^\beta \text{ and } V^\sigma = 0. \tag{3.3}$$

Thus if the intensity of $B$ is $b$, where

$$b = B/V,$$

and as an example, the intensity of amount of substance $i$ is the concentration of $i$

$$c_i = n_i/V$$

then

$$B_{model} = b^\alpha V^\alpha + b^\beta V^\beta \tag{3.4}$$

(a)

Phase $\beta$

Phase $\alpha$

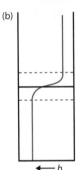

(b)

$\longleftarrow b$

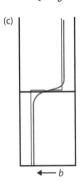

(c)

$\longleftarrow b$

**Figure 3.1.** Schematic, showing models of the interface: (a) represents the system; (b) shows the surface phase approach; (c) shows the surface excess approach, where the blue line represents the actual values of $b$ and the gray line (slightly offset) represents the model.

and for the example

$$n_{i\,\mathrm{model}} = c_i^\alpha V^\alpha + c_i^\beta V^\beta. \tag{3.5}$$

Consequently the surface excess quantity is

$$B^\sigma = B_{\mathrm{real}} - (b^\alpha V^\alpha + b^\beta V^\beta)$$

and for the example

$$n_i^\sigma = n_i - (c_i^\alpha V^\alpha + c_i^\beta V^\beta). \tag{3.6}$$

This model is illustrated in Figure 3.1(c), where the blue line shows the real values of $b$ and the gray line the model.

In the discussion that follows, the surface excess properties approach has been used. This treatment was extensively developed by Gibbs in the nineteenth century and is the basis of most present day surface thermodynamics, although the less popular and theoretically more complex surface phase approach also yields valid results.

## 3.3 Adsorption

The example used above relating to the amount of substance $i$ (Eqs. 3.4 and 3.6) was chosen because it is central to the concept of adsorption.

### 3.3.1 Adsorption and the dividing surface

In Eq. (3.6), $n_i^\sigma$ is the surface excess of $i$ at the total area, $A$, of the interface and is often called the total adsorption of $i$. For unit area of interface the term adsorption of $i$ ($\Gamma_i$) is used:

$$\Gamma_i = n_i^\sigma / A. \tag{3.7}$$

The value of $\Gamma_i$ (and $n_i^\sigma$) depends on the location of the mathematical dividing surface. Figure 3.2 is a similar representation to Figure 3.1, except that for clarity the distance is plotted on the $x$-axis. The area under the blue curve is

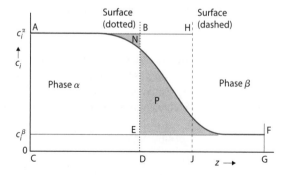

**Figure 3.2.** Plots of the concentration of $i$ against distance from the interface for the real (full blue line) and model (dashed and dotted black lines) interfaces. Adsorption of $i$ using the dotted surface BD is given by the shaded area P (where the real values are greater than the model values) minus the shaded area N (where the real values are less than those of the model). Note that the positions of the vertical lines AC and FG are arbitrary.

proportional to the actual number of moles of substance $i$. Consider firstly the dividing surface represented in two dimensions by the dotted line BD. For this model the number of moles of $i$ in phase $\alpha$ is proportional to the area of the rectangle ABDC. The number of moles in phase $\beta$ in this model is proportional to the area of the rectangle DEFG. Notice that they are very different to the actual amounts for each phase, which are proportional to the areas under the blue line on either side of the line BD. The differences between the real system and the model system are shaded and the surface excess, $n_i^\sigma$, is the shaded area P minus the shaded area N.

It is also apparent from Figure 3.2 that a change in the position of the surface affects the surface excess (and hence adsorption). In this example there is positive adsorption of $i$ if the dividing surface BD is selected, but negative adsorption if the dashed surface HJ is used. If the surface is placed in an intermediate position such that the shaded areas are equal, the adsorption of $i$ would be zero.

It is therefore necessary to specify the position of this surface, but as the transition from $c_i^\alpha$ to $c_i^\beta$ may extend over only a few molecular dimensions it must be done with greater accuracy than any physical measurement would permit. The position of the interface is therefore set by using a suitable convention.

### 3.3.2 Conventions for the dividing surface

Any convenient convention may be adopted to define the position of the dividing surface, but as the value of the adsorption will depend on the convention used, that convention must be specified. For example, the concentration of ethanol at the air–aqueous solution interface is higher than its concentration in the bulk solution phase, but there are, broadly, two ways of describing this situation: we may say that ethanol is positively adsorbed, or that water is negatively adsorbed.

For locating the position of the dividing surface there are two conventions in common use.

#### Gibbs convention

Gibbs, in the nineteenth century, suggested that a major component (A) should be selected and the dividing surface positioned so that the adsorption of A is zero:

$$\Gamma_A = 0. \tag{3.8}$$

For the other components the adsorptions are usually not zero.

In the case of ethanol + water mixtures, water could be chosen as A leading to positive adsorption values for ethanol (B), but equally well, ethanol could be chosen as A leading to negative adsorption values for water.

---

Josiah Willard Gibbs

Josiah Willard Gibbs (1839–1903) made outstanding contributions to physical science, and particularly to thermodynamics. One of his interests was surface phenomena and his work forms the basis of much of present day surface thermodynamics. His scientific writings were collected and published as a two-volume edition, *Collected Works of J. Willard Gibbs*, in 1928.

---

**Relative adsorption**

This convention relates adsorptions to the adsorption of one major component (A). For components A and $i$, Eqs. (3.6) and (3.3), give:

$$n_A^\sigma = n_A - c_A^\alpha V + (c_A^\alpha - c_A^\beta)V^\beta \tag{3.9}$$

$$n_i^\sigma = n_i - c_i^\alpha V + (c_i^\alpha - c_i^\beta)V^\beta. \tag{3.10}$$

Of the quantities on the right of Eqs. (3.9) and (3.10), only $V^\beta$ depends on the position of the dividing surface. It can be eliminated by combining the two equations:

$$n_i^\sigma - n_A^\sigma \left(\frac{c_i^\alpha - c_i^\beta}{c_A^\alpha - c_A^\beta}\right) = n_i - c_i^\alpha V - (n_A - c_A^\alpha V)\left(\frac{c_i^\alpha - c_i^\beta}{c_A^\alpha - c_A^\beta}\right).$$

In this equation, $n_i^\sigma$ and $n_A^\sigma$ are still dependent on the position of the dividing surface, but all terms on the right are independent of this choice. Thus the left side as a whole is also independent of the choice. This quantity, divided by the surface area, is called the relative adsorption of $i$ with respect to A:

$$\Gamma_i^A = \frac{1}{A}\left[n_i - c_i^\alpha V - (n_A - c_A^\alpha V)\left(\frac{c_i^\alpha - c_i^\beta}{c_A^\alpha - c_A^\beta}\right)\right]$$

$$= \Gamma_i - \Gamma_A\left(\frac{c_i^\alpha - c_i^\beta}{c_A^\alpha - c_A^\beta}\right). \tag{3.11}$$

Thus $\Gamma_i^A$ does not depend on the position of the dividing surface, and $\Gamma_i$ and $\Gamma_A$ may be defined by any convention without affecting $\Gamma_i^A$. If the Gibbs convention is used:

$$\Gamma_A(\text{Gibbs}) = 0$$

and hence

$$\Gamma_i(\text{Gibbs}) = \Gamma_i^A. \tag{3.12}$$

The relative adsorption of $i$ with respect to A is therefore equal to the adsorption of $i$ using the Gibbs convention with the adsorption of A set at zero. The utility of the Gibbs convention arises chiefly from this equality.

Other conventions have also been proposed and for certain special situations they may be more useful than either of the above conventions.

## ✗ 3.3.3 Adsorption isotherms

Adsorption data are often presented as an adsorption isotherm. This is a graph of the adsorption at a constant temperature plotted against some measure of the equilibrium bulk phase concentration of the adsorbing substance (the *adsorbate*). Here, *equilibrium* refers to the final bulk phase concentration after adsorption equilibrium has been established. The concentration may be expressed as the pressure for a gaseous adsorbate or the concentration of a solute adsorbate in a solution. Examples will appear in later chapters.

Equations describing adsorption isotherms have been developed for most situations and will be described as isotherm equations. Some of these equations are empirical, but others have been developed from models of the adsorption process or adsorption equilibrium. However the observation that an equation developed from a particular model gives a good fit to the experimental isotherm is not generally sufficient to validate the model: agreement in other aspects of the adsorption, especially thermal effects, is desirable.

## 3.4 Thermodynamic properties of interfaces

The total adsorption, $n_i^\sigma$, is just one example of a surface excess quantity. Other extensive properties can be treated in a similar way, so that we may have, for example, a surface excess free energy, a surface excess entropy, a surface excess internal energy and so on. We note however that for a system in equilibrium, the temperature is uniform throughout and that although the pressures in the two bulk phases may be different (see Section 2.7) there is no surface excess pressure as this quantity is effectively replaced by the surface tension. We recall also that the surface excess volume is zero.

To develop expressions for these various surface excess properties, we begin by enlarging the concept of mechanical work to include the work associated with interfaces within the system and then incorporate this into the first law of thermodynamics.

### 3.4.1 Mechanical work

For a system containing an interface the exchange of mechanical work with the surroundings can take place by the usual mechanism of volume change or by a change in the interfacial area. Thus for mechanical work, $w_{\text{mech}}$

$$\delta w_{\text{mech}} = -P^\alpha \, dV^\alpha - P^\beta \, dV^\beta + \gamma \, dA \tag{3.13}$$

where the first two terms allow for the difference in bulk pressure when the interface is curved and the third term comes from Eq. (3.2). Furthermore, as $P^\alpha$ and $P^\beta$ are internal pressures this relationship is only valid for reversible processes. This implies that either $P^\alpha$ or $P^\beta$ is in near equilibrium with the external pressure.

If the system is open, it is usual to include terms for the exchange of matter with the surroundings in the expression for work, giving

$$\delta w = -P^\alpha \, dV^\alpha - P^\beta \, dV^\beta + \gamma \, dA + \sum_i \mu_i \, dn_i \tag{3.14}$$

where the summation in the last term extends over all components in the system.

### 3.4.2 Internal energy and enthalpy

When this expression for work is introduced into the usual statement of the first law of thermodynamics, we obtain

$$dU = \delta q + \delta w$$
$$= T \, dS - P^\alpha \, dV^\alpha - P^\beta \, dV^\beta + \gamma \, dA + \sum_i \mu_i \, dn_i \tag{3.15}$$

for a reversible process.

For this system we can define the enthalpy as

$$H = U + P^{\alpha} V^{\alpha} + P^{\beta} V^{\beta}. \tag{3.16}$$

We note that another definition which includes the surface term ($\gamma A$) is also possible and is more useful in certain conditions.

Differentiating (3.16) and substituting from (3.15) yields

$$dH = T \, dS + V^{\alpha} \, dP^{\alpha} + V^{\beta} \, dP^{\beta} + \gamma \, dA + \sum_i \mu_i \, dn_i. \tag{3.17}$$

### 3.4.3 Gibbs free energy

The usual definition of Gibbs free energy ($G$)

$$G = H - TS$$

now gives

$$dG = -S \, dT + V^{\alpha} \, dP^{\alpha} + V^{\beta} \, dP^{\beta} + \gamma \, dA + \sum_i \mu_i \, dn_i. \tag{3.18}$$

For a process that involves only a change in interfacial area

$$\left( \frac{\partial G}{\partial A} \right)_{T, P^{\alpha}, P^{\beta}, n_i} = \gamma. \tag{3.19}$$

Equation (3.19) provides a convenient thermodynamic definition of the surface tension.

Note that if the alternative definition of enthalpy is adopted the definition of Gibbs free energy also changes.

## 3.5 Surface excess quantities

Surface excess quantities may be manipulated in the same way as other thermodynamic quantities. Take Gibbs free energy as an example:

$$dG = -S \, dT + V \, dP + \gamma \, dA + \sum_i \mu_i \, dn_i. \tag{3.20}$$

Each of the extensive properties can be expanded to show the contributions from the bulk phases and the interface, as in Eq. (3.1):

$$
\begin{aligned}
dG ={}& dG^{\alpha} + dG^{\beta} + dG^{\sigma} \\
={}& -(S^{\alpha} + S^{\beta} + S^{\sigma})dT + V^{\alpha}dP^{\alpha} + V^{\beta}dP^{\beta} + \gamma dA \\
& + \sum_i \mu_i^{\alpha} \, dn_i^{\alpha} + \sum_i \mu_i^{\beta} \, dn_i^{\beta} + \sum_i \mu_i^{\sigma} \, dn_i^{\sigma}.
\end{aligned} \tag{3.21}
$$

For the bulk phases

$$dG^{\alpha} = -S^{\alpha} \, dT + V^{\alpha} \, dP^{\alpha} + \sum_i \mu_i^{\alpha} \, dn_i^{\alpha} \tag{3.22}$$

$$dG^{\beta} = -S^{\beta} \, dT + V^{\beta} \, dP^{\beta} + \sum_i \mu_i^{\beta} \, dn_i^{\beta} \tag{3.23}$$

and subtraction of (3.22) and (3.23) from (3.21) gives

$$dG^\sigma = -S^\sigma dT + \gamma dA + \sum_i \mu_i^\sigma dn_i^\sigma. \tag{3.24}$$

Note the similarity between (3.24) and (3.22)/(3.23): the *pressure–volume* term has been replaced by the *surface-tension–area* term.

At constant temperature, integration of (3.24) gives

$$\frac{G^\sigma}{A} = \gamma + \sum_i \mu_i^\sigma \Gamma_i. \tag{3.25}$$

Similarly,

$$dU^\sigma = dH^\sigma = T\,dS^\sigma + \gamma\,dA + \sum_i \mu_i^\sigma\,dn_i^\sigma \tag{3.26}$$

and the integrated forms give

$$\frac{U^\sigma}{A} = \frac{H^\sigma}{A} = \gamma + \frac{TS^\sigma}{A} + \sum_i \mu_i^\sigma \Gamma_i. \tag{3.27}$$

Note that if the system is in equilibrium, the chemical potential of $i$ will be the same in all three regions $\alpha$, $\beta$, and $\sigma$, and the superscripts on $\mu_i$ may be omitted.

## 3.6 Measurement of adsorption

Methods for measuring adsorption can be grouped into three categories: concentration change, surface analysis, and surface tension change.

 ### 3.6.1 Adsorption from concentration change

This method is used for adsorption on solids with large surface area, as the change in adsorbate concentration due to adsorption must be large enough to measure.

For such a measurement we might take a solution of solute $i$ and add to it a solid adsorbent. The initial concentration of $i$ is denoted by $c_i^\circ$ and the final equilibrium concentration after adsorption is $c_i$. $V$ is the volume of solution, and $A$ the area of solid surface.

Total amount of $i$ present:       $n_i^\circ = c_i^\circ V$.
Amount in solution after adsorption:    $n_i = c_i V$.
The amount lost from the solution is the total amount adsorbed:

$$n_i^\sigma = n_i^\circ - n_i = \left(c_i^\circ - c_i\right) V. \tag{3.28}$$

Equation (3.28) can then be used to calculate $\Gamma_i$ if $A$ is known, or to calculate $A$ if $\Gamma_i$ is known.

Examples

Adsorption of a gas on a solid (see Chapter 8) and the adsorption of acetic acid (from aqueous solution) on charcoal (see Section 9.3.)

### 3.6.2 Adsorption from surface analysis

This approach can sometimes be applied to fluid interfaces, but is not often used. The method involves sampling the liquid in the interfacial region and analysing both it and the bulk liquid. For the adsorption of solute $i$ at the gas–solution interface, the total amount of $i$ is:

$$n_{i\ \mathrm{real}} = c_i^{\mathrm{L}} V^{\mathrm{L}} + c_i^{\mathrm{S}} V^{\mathrm{S}} + c_i^{\mathrm{G}} V^{\mathrm{G}},$$

where L, G, S refer to liquid, gas, and surface sample, respectively. This is the 'real' system. For the 'model' system there is no adsorption and the concentration of $i$ in the surface sample is the same as in the bulk liquid. Thus:

$$n_{i\ \mathrm{model}} = c_i^{\mathrm{L}} V^{\mathrm{L}} + c_i^{\mathrm{L}} V^{\mathrm{S}} + c_i^{\mathrm{G}} V^{\mathrm{G}}.$$

From Eq. (3.2) the total amount of $i$ adsorbed is:

$$n_i^{\sigma} = n_{i\ \mathrm{real}} - n_{i\ \mathrm{model}} = \left(c_i^{\mathrm{S}} - c_i^{\mathrm{L}}\right) V^{\mathrm{S}}. \tag{3.29}$$

Examples

Foam collection, surface microtome, radio-labelled solutes, low-angle neutron and X-ray reflection.

### 3.6.3 Adsorption from surface tension change – the Gibbs adsorption isotherm

Adsorption at a fluid interface causes a change in the surface tension which can be used to determine the extent of adsorption. The relationship is given by the Gibbs adsorption isotherm or simply the Gibbs equation:

$$\frac{-\mathrm{d}\gamma}{RT} = \sum_i \left(\Gamma_i \mathrm{d}\ln(a_i)\right). \tag{3.30}$$

The summation must be taken over all species present in the solution, but usually some terms are zero or negligibly small. The Gibbs convention, Eq. (3.8), means that the term for the solvent is zero. For a two-component system and using the Gibbs convention:

$$\frac{-\mathrm{d}\gamma}{RT} = \Gamma_{\mathrm{B}} \mathrm{d}\ln(a_{\mathrm{B}}) \tag{3.31}$$

or

$$\Gamma_{\mathrm{B}} = -\frac{1}{RT} \frac{\mathrm{d}\gamma}{\mathrm{d}\ln(a_{\mathrm{B}})} = \frac{a_{\mathrm{B}}}{RT} \frac{\mathrm{d}\gamma}{\mathrm{d}(a_{\mathrm{B}})}. \tag{3.32}$$

The technique thus involves measuring the surface tensions of a set of solutions with different concentrations to find the dependence of surface tension on concentration. At any given concentration the slope of the plot of $\gamma$ against

---

### Derivation of the Gibbs equation

We begin with the expression for the change in surface excess internal energy derived above as Eq. (3.26):

$$dU^\sigma = T\,dS^\sigma + \gamma\,dA + \sum \mu_i\,dn_i^\sigma. \qquad (3.26), (3.30a)$$

Integration of (3.30a) under conditions of constant $T$, $\gamma$ and $\mu$ gives

$$U^\sigma = TS^\sigma + \gamma A + \sum \mu_i n_i^\sigma. \qquad (3.30b)$$

Since $U$ is a state function, this holds under all conditions, and we can take the total differential

$$dU^\sigma = T\,dS^\sigma + S^\sigma\,dT + \gamma\,dA + A\,d\gamma + \sum \mu_i\,dn_i^\sigma + \sum n_i^\sigma\,d\mu_i. \qquad (3.30c)$$

Comparing equations (3.30a) and (3.30c) gives a Gibbs–Duhem type equation:

$$S^\sigma\,dT + A\,d\gamma + \sum n_i^\sigma\,d\mu_i = 0$$

which rearranges to the Gibbs adsorption equation:

$$d\gamma = -(S^\sigma/A)dT - \sum \Gamma_i\,d\mu_i. \qquad (3.30d)$$

The chemical potential of substance $i$ is

$$\mu_i = \mu_i^\theta + RT \ln a_i$$

and at constant temperature

$$d\mu_i = RT\,d(\ln a_i)$$

so under these conditions, substitution into (3.30d) gives

$$-\frac{d\gamma}{RT} = \sum \left(\Gamma_i\,d(\ln a_i)\right) \qquad (3.30)$$

where the summation extends over all chemical species in the solution.

---

concentration or ln (concentration) leads to the value of adsorption under those conditions.

## 3.7 Adsorption from solutions

In using the Gibbs adsorption equation (3.30) it is essential to consider carefully the nature of the solute. For example, the solute may be ionized in solution or aggregates may form in addition to the individual molecules and this means that there are two, or possibly more, solute species present. Another potential source of error is partial ionization of the solute. These and other complications will be discussed in Section 4.6.3.

For simple unionized solutes, the form of the Gibbs equation in Eq. (3.32) is applicable. If the solution is dilute, activity may be replaced by concentration:

$$\Gamma_B = -\frac{1}{RT}\frac{d\gamma}{d\ln(c_B)} = -\frac{c_B}{RT}\frac{d\gamma}{d(c_B)}. \qquad (3.33)$$

## SUMMARY

Two models of the interface are presented and the surface excess model is selected for further development. In this model the total value of a property for the real system is compared with the value for a model system in which the interface is a mathematical surface and the intensive properties of the bulk phases extend unchanged right up to that surface. The difference between the total values of a property for the real and the model systems gives the surface excess value of that property. The surface excess amount of a component material for unit area of surface is called the adsorption.

This approach requires the mathematical dividing surface to be positioned with an accuracy that is not possible by physical measurement so a convention is adopted. The most commonly used convention is that of Gibbs which positions the surface to make the adsorption of one major component equal to zero. All other surface excess properties for all other component substances are then referred to this surface.

Techniques for measuring adsorption can be grouped into three classes: concentration change, surface analysis, and surface tension change. For the last of these, the Gibbs adsorption equation is developed and discussed.

## FURTHER READING

Adamson, A. W. and Gast, A. P. (1997). *Physical Chemistry of Surfaces, 6th edn.* Wiley-Interscience, New York.

Defay, R., Prigogine, I., Bellemans, A., and Everett, D. H. (1966). *Surface Tension and Adsorption.* Longmans, London.

Hunter, R. J. (2001). *Foundations of Colloid Science, 2nd edn.* Oxford University Press, Oxford.

## EXERCISES

**3.1.** A $0.5\,cm^3$ sample, collected from the surface of a surfactant solution, is found to contain $3.5\,\mu mol$ of surfactant. The parent solution $(80\,cm^3)$ has a concentration of $4.00\,mol\,m^{-3}$. Calculate the total amount of surfactant adsorbed at the surface.

**3.2.** From surface tension measurements on aqueous solutions of butanol, it was found that the slope of the graph of surface tension against concentration was $-0.156\,mN\,m^2\,mol^{-1}$ at a bulk concentration of $6.40\,mol\,m^{-3}$. Calculate the adsorption at this concentration and state any assumptions made.

**3.3.** An aqueous solution of acetic acid $(50\,cm^3)$ was shaken with charcoal $(2.5\,g)$ until equilibrium was reached. The concentration of the original solution was $0.010\,mol\,dm^{-3}$, while after adsorption the concentration had fallen to $0.006\,mol\,dm^{-3}$. Calculate the amount of acetic acid adsorbed on $1.0\,g$ of charcoal.

**3.4.** From the following surface tension data for the ethanol + water system, calculate (a) the adsorption of ethanol at mole fraction 0.8, treating water as the solvent in the Gibbs convention; (b) the adsorption of water at the same concentration, treating ethanol as the solvent. $(T = 298\,K)$.

| Concentration, $x$(ethanol) | Surface tension, $\gamma/\mathrm{mN\,m^{-1}}$ |
|---|---|
| 1.00 | 22.31 |
| 0.89 | 22.75 |
| 0.78 | 23.23 |
| 0.70 | 23.78 |
| 0.61 | 24.32 |
| 0.48 | 25.48 |

**3.5.** Derive expressions for $\mathrm{d}U^{\sigma}$ and $\mathrm{d}H^{\sigma}$ and show that these two quantities are identical (see Eq. 3.26).

# 4 Adsorption at the gas–liquid interface

## 4.1 Introduction

Some of the most obvious and dramatic effects of adsorption can be seen at the air–solution interface. Liquid drops change their shape, foams can be formed on liquids that otherwise will not foam, the rise of liquids in capillaries is reduced, the spreading of liquids over solid surfaces is enhanced, the ability of light winds to ruffle the surface of water in a pond is reduced, soils are wetted more readily by water, and the cleaning of materials is improved.

## 4.2 Measurement of equilibrium adsorption

As the area of a gas–liquid interface is usually small and sample collection is often difficult, the measurement of adsorption by analysis of the surface or by change in bulk phase concentration is usually not feasible. Thus adsorption is normally determined from surface tension data using the Gibbs equation (Eq. 3.30), but there can be complications in the chemistry of the solutions that may require additional information or the use of other methods where that is possible. Some of these problems will be discussed later in this chapter.

Generally, as the concentration of the solute is increased the surface tension of the solution decreases, indicating that the solute species is being positively adsorbed at the interface. There are however a few solutes that are negatively adsorbed (i.e. they are excluded from the interface) and then the surface tension rises with increasing concentration. Such rises are generally quite small, which is understandable as the composition of the surface does not change very much from that of the pure solvent (see Table 4.1).

### 4.2.1 Surface tension measurement

Techniques for measuring surface tension have been described in Section 2.6. Capillary rise is a simple and accurate technique, but cleanliness of the capillary is critical and for accurate work corrections for the non-sphericity of the meniscus must be made. There is also the possibility that the solute may contaminate the capillary surface so this technique is unsuitable for surfactants that adsorb on glass (such as $C_{16}TAB$).

A quick and reasonably accurate technique is the detachment of a paper Wilhelmy plate. With the plate attached to a sensor such as a load cell or strain gauge the liquid can be repeatedly raised to contact the plate and lowered to detach from the plate with comfortable speed (say, one cycle every 3–4 s). The difference between the maximum and minimum load readings is the difference between the force on the plate at detachment and in the air. This difference gives the surface tension. Sources of error include neglect of the negative buoyancy as the plate detaches from the meniscus it has raised and the weight of a rim of liquid that usually remains on the bottom edge of the plate after detachment. To some extent these errors cancel and the results are sufficiently accurate for most purposes.

### 4.2.2 Surface analysis

Methods of surface analysis are those where the amount of solute at the surface is measured directly and compared to that in the bulk.

Radio-labelled solutes can be detected at the surface by, for example, a Geiger tube, but calibration can be a problem. There can also be considerable difficulty in obtaining some labelled solutes of sufficiently high purity for accurate work.

Neutron reflectometry using deuterated solutes gives a measure of the amount of solute at the surface. Again, there can be difficulty in obtaining deuterated solutes of sufficient purity.

Because X-ray reflectometry measurements do not require specially prepared materials (deuterated or radio-labelled), highly purified solutes are more readily available. Moreover, the generally higher intensity of X-ray sources relative to neutron sources results in better reflectivity data. However the applicability of X-ray reflectivity for adsorption measurement is more limited than that of neutron reflectivity because it is not always possible to ensure that the signal results only from the material of interest. Further details of the neutron and X-ray reflectivity techniques are given in Section 5.5.

For some surfactants the formation and collection of a foam enables the measurement of adsorption by surface analysis (Weil, 1966). A measured quantity of an inert gas is bubbled through the solution forming a foam which passes up a long tube to drain and then through an inverted U-bend to a collection vessel where the foam is collapsed by cooling (or heating). The volume and concentration of the collapsed foam are measured and also the concentration of the parent solution. These data give the total amount adsorbed. To calculate the adsorption per unit area of the gas–solution interface it is necessary to estimate the area of foam surface. This is the major source of error with the technique. Reasonably uniform bubbles can be generated by careful design of the bubbler orifice and bubble size can be measured by passing some of the foam through a calibrated capillary. These results, together with the volume of gas used, yield the total surface area. The technique can only be used with solutes and at concentrations that are capable of producing a stable foam that will move smoothly up a glass chimney.

## 4.3 Observation of adsorption kinetics

Non-equilibrium situations arise in many practical applications, particularly those involving surfactants, so the study of adsorption kinetics assumes considerable importance. Experimental techniques can be grouped into two classes: those where fresh surface is created and the rate of attainment of equilibrium is measured; and those where an established equilibrium is disturbed.

### 4.3.1 Adsorption at freshly formed interfaces

An example of this approach is the study of jets of solution emerging from a nozzle. At the orifice the surface of the liquid would have the same composition as the bulk liquid, i.e. there would be zero adsorption, but adsorption would begin immediately the liquid emerges and could be monitored by the consequent changes in interfacial tension. One way of following this process is to use an oscillating jet. If the orifice is oval in shape the emerging jet will be oval in cross-section, but the forces of interfacial tension will tend to make it circular. Usually overshoot will occur and an oval cross-section will again be formed but with the long axis at right angles to the original long axis. This process will continue for some distance along the jet. Clearly the higher the surface tension the stronger and more rapid the oscillations will be, so measurements of the length (and hence the time) of each oscillation show how the interfacial tension changes along the jet and give the rate of adsorption. For a pure liquid the nodes will be equally spaced, but the adsorption of a solute will cause the distance between nodes to increase along the jet.

Changes in the adsorption rate arising from the oscillation of the jet are avoided with a circular jet. In this case the adsorption along the jet can be measured by changes in the surface potential (see Section 5.5.2).

For somewhat slower adsorption processes, the overflow of solution from a circular tube may be used. The apparatus consists of a vertical glass tube with a ground upper edge so that it is easily wetted, a collection vessel for the overflow, a pump, and a sintered glass disk in the inner tube to dampen currents (Figure 4.1). Surface tension is measured by a static Wilhelmy plate.

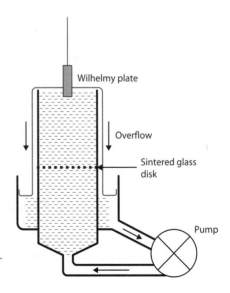

Figure 4.1. Cross-section of equipment for measuring rates of adsorption at the air–solution interface.

The age of surface detected by the Wilhelmy plate varies with the pumping rate, but because the surface age is averaged over the length of the plate an absolute calibration is not possible. Thus only trends can be observed.

### 4.3.2 Surface waves

Both transverse capillary waves and longitudinal waves can deliver information about the elasticity and viscosity of surfaces, albeit on very different time-scales. Rates of adsorption and desorption can also be deduced.

**Transverse capillary waves** are usually generated with frequencies between 100 and 300 Hz. The generator is a hydrophobic knife edge situated in the surface and oscillating vertically, while the usual detector is a light-weight hydrophobic wire lying in the surface parallel to the generator edge. The detector is connected to a transducer that converts the vertical motions into an electrical signal. The generator and detector are usually close together (15–20 mm) so reflections set up a pattern of standing transverse waves. Optical detection, on the other hand, causes no interference to the generated wave pattern.

The damping of capillary ripples arises primarily from the compression and expansion of the surface and the interaction between surface film and underlying liquid. In a transverse progressive wave, volume elements of the liquid tend to move in circular paths with motion in the direction of propagation at the top of the path and in the reverse direction at the bottom (Figure 4.2). This leads to compression and expansion of the surface as the wave passes.

If a surface film is present, compression tends to lower the surface tension while expansion raises it. This generates a Marangoni flow (see Section 4.10) which opposes the wave motion and dampens it. Furthermore, if the film material is soluble, the compression–expansion cycle will be accompanied by a desorption–adsorption cycle. The dissipation of energy associated with these processes contributes to the damping of waves. It explains such phenomena as the calming effect when 'oil is poured on troubled waters'. Reviews of the hydrodynamic theories for capillary ripples have been presented by Lucassen-Reynders and Lucassen (1969) and Hansen and Ahmad (1971).

For **longitudinal waves** a barrier (usually on the trough of a surface film balance – see Section 5.2.3) is moved horizontally to generate small-amplitude sinusoidal oscillations in area. At frequencies less than about 0.1 Hz the wavelength is large compared to the (usual) length of the trough and reflections from the ends of the trough then combine so that the entire surface is deformed uniformly and simultaneously. Measurement of the changes in surface tension at any convenient location then gives the *surface*

**Figure 4.2.** Motions of fluid elements in capillary ripples. The surface expands on the left of the crest and is compressed on the right.

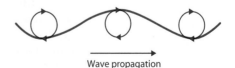

Wave propagation

*dilational elasticity and viscosity* (see Section 5.5.3, Eq. 5.6). A phase difference between the area change and the change in surface tension indicates a relaxation process which could arise from:

- movement of solute to and from the bulk liquid phase to the subsurface region;
- movement to and from the subsurface region to the surface (adsorption–desorption);
- rearrangement of the material of the adsorbed film towards its instantaneous equilibrium state.

For small molecules the rearrangement process is very fast compared to the oscillation frequency and so does not affect the measured viscosity. Similarly, it is often assumed that the establishment of equilibrium between the surface film and the subsurface region is rapid, leaving only the movement of solute between bulk liquid and the subsurface as the source of visco-elastic behaviour. Such behaviour has been treated by Lucassen and van den Tempel (1972).

For longitudinal waves of higher frequency the situation is more complex. At frequencies above about 10 Hz, the compression waves are damped out before they reach the end of the trough so reflection and interference are negligible. A time delay and a decay in amplitude between barrier movement and the consequent change in surface tension at the point of measurement must be factored into the analysis. For intermediate frequencies the changes in surface tension vary in a complex pattern along the trough, but in most cases this problem can be circumvented if measurements are taken a small distance before the mid-length of the trough where the amplitude and phase have been shown to be independent of frequency (Lucassen and Barnes, 1972).

## 4.4 Adsorption of non-electrolyte solutes

### 4.4.1 Effects of solute concentration on the surface tension

A graph of surface tension against solute concentration often has the form shown in Figure 4.3, the surface tension and the negative slope both decreasing as the concentration increases.

The slope, $d\gamma/dc_B$, can be determined from such curves, particularly if, as in Figure 4.3, the data can be fitted by a function that can be readily differentiated. Sometimes the plot of surface tension against $\ln c_B$ is nearly linear, enabling a more accurate determination of the slope to be made. This is not the case for the data in Figure 4.3.

Once the slope, $d\gamma/dc$ or $d\gamma/d\ln c$, has been determined the adsorption can be calculated from the Gibbs equation (3.32) as shown in Figure 4.4.

This adsorption isotherm may be described as a *composite* adsorption isotherm. The theory behind this concept is discussed in detail in Section 9.4.1. The decrease in the adsorption of butanol after the peak is due to the increasing amount of non-adsorbed butanol at the surface (i.e. the

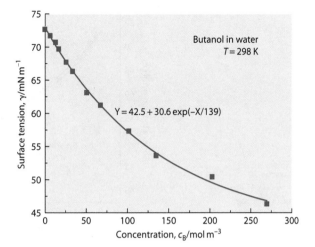

**Figure 4.3.** Effect of butanol concentration on the surface tension of an aqueous solution (data from Posner *et al.*, 1952).

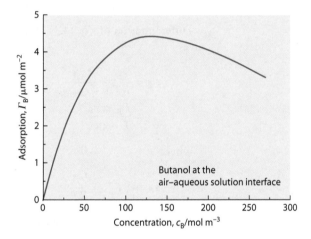

**Figure 4.4.** Adsorption of butanol at the surface of butanol + water solutions. Adsorption values are calculated from the data in Figure 4.3 using the Gibbs equation.

amount that would be at the surface if there were no adsorption) and competition for the remaining sites from the water. Note that as we approach pure butanol the *adsorption* falls to zero even though the surface layer is pure butanol.

### 4.4.2 Szyszkowski–Langmuir adsorption

For many binary systems, A + B, the surface tension follows the Szyszkowski relation:

$$\gamma = \gamma_A^\bullet - b\gamma_A^\bullet \ln(1 + c_B/a) \tag{4.1}$$

where $a$, $b$ are constants and $\gamma_A^\bullet$ represents the surface tension of pure A.

Figure 4.5 shows that the butanol-in-water system follows the Szyszkowski relation at low concentrations, but deviates significantly at higher concentrations, a trend that can be attributed to competition from the solvent water.

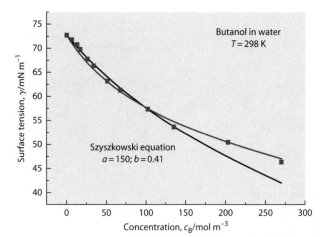

**Figure 4.5.** Fit of the Szyszkowski equation ($a = 150\,\mathrm{mol\,m^{-3}}$, $b = 0.41$) to experimental data (■) for the butanol + water system.

Differentiation of Eq. (4.1) and substitution in the Gibbs equation (3.32) gives

$$
\begin{aligned}
\Gamma_{\mathrm{B}} &= \frac{c_{\mathrm{B}}}{RT} \frac{b\gamma_{\mathrm{A}}^{\bullet}/a}{1 + c_{\mathrm{B}}/a} \\
&= \frac{\Gamma_{\mathrm{B}}^{\infty}\,\alpha\,c_{\mathrm{B}}}{1 + \alpha\,c_{\mathrm{B}}}
\end{aligned}
\tag{4.2}
$$

where

$$
\alpha = \frac{1}{a}; \; \Gamma_{\mathrm{B}}^{\infty} = \frac{b\gamma_{\mathrm{A}}^{\bullet}}{RT}.
$$

A form suitable for linear plotting can be obtained:

$$
\frac{c_{\mathrm{B}}}{\Gamma_{\mathrm{B}}} = \frac{1}{\Gamma_{\mathrm{B}}^{\infty}\alpha} + \frac{c_{\mathrm{B}}}{\Gamma_{\mathrm{B}}^{\infty}}.
\tag{4.3}
$$

Equation (4.2) has the same form as the Langmuir adsorption isotherm, originally derived for the chemisorption of gases on solids and discussed in detail in Section 8.5.1. Equations (4.2) and (4.3) are sometimes called the Szyszkowski–Langmuir isotherm.

Again it is only the data at low concentrations that fit the Szyszkowski–Langmuir isotherm.

According to the Langmuir equation, at high concentrations $\Gamma_{\mathrm{B}} \to \Gamma_{\mathrm{B}}^{\infty}$, so the adsorption tends to a limiting value. The Langmuir equation was originally derived for adsorption at the gas–solid interface in which case the limiting behaviour corresponds to complete monolayer coverage. At the gas–liquid interface the limit is generally lower and may not be reached because of competition from the solvent (as in Figure 4.4). With butanol in water, for example, $\Gamma_{\mathrm{B}}^{\infty}$ is significantly lower than theoretical upper limit of about $8.3\,\mu\mathrm{mol\,m^{-2}}$ (see Figure 4.6).

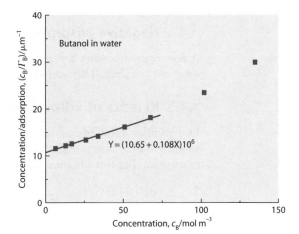

**Figure 4.6.** Test of Eq. (4.3) for the data in Figure 4.4. Note that only the data at low concentrations fit the straight line. From this line we obtain $\Gamma_B^{\infty} = 5.55\,\mu\text{mol m}^{-2}$ $(0.30\,\text{nm}^2\,\text{molecule}^{-1})$, $\alpha = 0.0101\,\text{mol}^{-1}\,\text{m}^2$.

### 4.4.3 Equations of state

The results of adsorption measurements may be expressed as an equation of state such as that formulated by Schofield and Rideal:

$$\Pi\left(\hat{A} - \hat{A}^{\circ}\right) = qkT \qquad (4.4)$$

or

$$\Pi\hat{A} = qkT + \Pi\hat{A}^{\circ} \qquad (4.5)$$

where $\Pi = \gamma_{\text{solvent}} - \gamma_{\text{solution}}$ is the surface pressure; $\hat{A}_i = 1/\Gamma_i$, is the area per molecule of $i$; and $q$ is a constant giving a measure of the affinity of the adsorbed molecules for each other. Generally, the amount adsorbed is less than one monolayer. Plots of Eq. (4.5) for various unionized fatty acids show good linear relationships over much of the surface pressure range. Some of these data are shown in Figure 4.7.

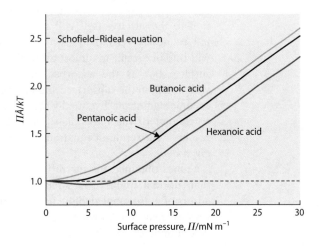

**Figure 4.7.** Tests of the Schofield–Rideal equation (4.5) for the adsorption of several unionized carboxylic acids at the air–aqueous solution interface. (Replotted from data of Schofield and Rideal, 1925.)

### 4.4.4 Negative adsorption

A few organic solutes are negatively adsorbed at the air–(dilute aqueous solution) interface. They include glycine and sucrose.

### 4.4.5 Kinetics of adsorption

In the initial stages of adsorption the rate is determined by diffusion of the solute to the surface, provided that there is no stirring and no energy barrier to adsorption. For this situation we can write:

$$\frac{d\Gamma}{dt} = \left(\frac{D}{\pi}\right)^{1/2} c t^{-1/2} \tag{4.6}$$

so that

$$\Gamma = 2c(Dt/\pi)^{1/2} \tag{4.7}$$

where $D$ is the diffusion coefficient and $c$ is the concentration of the solute.

Even though this equation only applies at the beginning of the adsorption process it does suggest that small amounts of impurity in the solute can take a long time to reach adsorption equilibrium and give rise to long-term drifts in surface tension. We will see in Section 4.7.6 that this effect can be used as a test for surfactant purity.

At later stages the back diffusion of solute away from the surface must also be considered. The net rate of adsorption is then slower than given by Eq. (4.6) and has been described by the theory of Ward and Tordai (1946). This theory also incorporates the possibility of an energy barrier to adsorption. Their model postulates a subsurface region with a thickness of only a few molecular diameters that is in instantaneous equilibrium with the surface. Effectively this enables the amount adsorbed at the surface to be expressed as a bulk-type concentration and thus simplifies the diffusion problem.

When a new surface is created the solute concentration in the subsurface region will be the same as that in the bulk, but nearly instantaneous adsorption to the surface itself will cause the concentration in the subsurface to fall to nearly zero. Diffusion of solute from the bulk to the subsurface will initially result in almost all of this material moving immediately to the surface, but as the adsorbed amount increases more and more solute will remain in the subsurface and back diffusion from subsurface to bulk will become significant. Eventually the solute concentration in the subsurface will reach that of the bulk phase and adsorption equilibrium will have been reached. The resultant equation requires graphical integration and will not be given here.

Values obtained for the diffusion coefficient that are significantly lower than normal (bulk) values indicate the existence of an energy barrier to adsorption. Iso-amyl alcohol, for example, gives a value that is about 56 times smaller than the normal value showing that diffusion is not the rate-determining step in the adsorption process.

## 4.5 Adsorption of ionized solutes

When the solute is ionized the Gibbs equation must be used in the general form with a term for each species in solution. Using the Gibbs convention and a solute that ionizes completely to $M^+$ and $X^-$;

$$\frac{-d\gamma}{RT} = \Gamma_{M^+} d\ln c_{M^+} + \Gamma_{X^-} d\ln c_{X^-}$$

If $M^+$ and $X^-$ are the only ions in the solution then electrical neutrality requires that

$$c_{M^+} = c_{X^-} = c \text{ and } \Gamma_{M^+} = \Gamma_{X^-} = \Gamma$$

and

$$-\frac{d\gamma}{RT} = 2\Gamma d\ln c$$

or

$$\Gamma = -\frac{1}{2RT}\frac{d\gamma}{d\ln c} = -\frac{c}{2RT}\frac{d\gamma}{dc}. \tag{4.8}$$

Aqueous solutions of simple electrolytes often show a slight rise in surface tension with concentration. According to the Gibbs equation (4.8) this indicates negative adsorption of the solute: in other words the electrolyte is partly expelled from the surface. Some typical values of surface tension are shown in Table 4.1.

**Table 4.1.** Surface tension values of some aqueous solutions of simple salts at $20\,^{\circ}C$ (data from Weast, 1977).

| Solute | Concentration, $c_B/\text{mol dm}^{-3}$ | Surface tension, $\gamma/\text{mN m}^{-1}$ |
|---|---|---|
| None | | 72.75 |
| NaCl | 0.10 | 72.92 |
| NaCl | 0.93 | 74.4 |
| NaCl | 4.43 | 82.55 |
| KCl | 0.93 | 74.15 |
| NaBr | 0.91 | 74.05 |
| NaNO$_3$ | 0.92 | 73.95 |
| BaCl$_2$ | 0.80 | 74.9 |
| MgCl$_2$ | 0.91 | 75.75 |
| Na$_2$CO$_3$ | 0.90 | 75.45 |
| HCl | 0.97 | 72.45 |
| H$_2$SO$_4$ | 0.84 | 72.55 |

## 4.6 Adsorption of surfactants

### 4.6.1 Molecular structure of surfactants

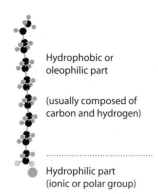

Hydrophobic or
oleophilic part

(usually composed of
carbon and hydrogen)

Hydrophilic part
(ionic or polar group)

**Figure 4.8.** Structural features
of a typical amphiphile (in this
case sodium hexadecanoate).

Surfactants (an abbreviation for *surface active agents*) belong to a class of compounds known as **amphiphiles**. An amphiphile (from Greek: *loves both*[2]) is a substance that combines in the one molecule a segment that tends to be water insoluble and oil soluble, i.e. hydrophobic or oleophilic, and a segment that tends to be water soluble, i.e. hydrophilic. The balance between the hydrophobic and hydrophilic segments can cover a wide range: from molecules that are almost completely insoluble in water to ones that are highly soluble. This is sometimes referred to as the hydrophilic–lipophilic balance or HLB (see Section 6.3 for more details). The essential features of a typical amphiphile are shown in Figure 4.8.

The surfactants are water-soluble amphiphiles and are strongly adsorbed at surfaces. The driving force for this strong adsorption comes from the amphiphilic nature of the surfactant molecule; for example, at the air–aqueous solution interface the hydrophobic portion of the molecule can be in the vapour phase out of the water while the hydrophilic portion can be immersed in the water (as in Figure 4.9). Such adsorbed films are known as Gibbs monolayers (*monolayer* because there is normally only one layer).

The use of a circle and line to represent a surfactant molecule is sometimes criticized because it does not adequately represent the relative sizes of the head group and the tail. Often the head group is actually smaller in cross-section than the tail. However, this is not the complete picture as the hydrophilic head group will have water molecules associated with it, increasing its effective size. Nevertheless it is useful to examine space-filling molecular structures to get a better perspective. Images of some such structures are shown in Figure 4.10.

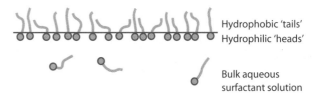

Hydrophobic 'tails'
Hydrophilic 'heads'

Bulk aqueous
surfactant solution

**Figure 4.9.** Schematic illustration of a Gibbs monolayer. It is a common convention to draw amphiphiles with a circular dot to represent the hydrophilic 'head' group and a line (irregular or straight) to represent the hydrophobic 'tail' section.

2. The terms *amphiphilic* and *amphipathic*, although opposite in literal meaning, apply to identical molecular types. The first term means *loves both*, the second means *hates both*. Thus, for example, a polar group 'likes' water but 'hates' non-polar solvents, an alkyl chain 'hates' water but 'loves' non-polar solvents. Until fairly recently the term *amphipathic* was normally used, but as love is always better than hate, *amphiphilic* is now preferred.

Many of the common amphiphiles have trivial names and some of these are given in Table 4.2 and Table 4.3.

There are three classes of surfactant, defined by the ionic nature of the surface active species: **anionic**, **cationic**, and **non-ionic**. Some examples are shown in Figure 4.11.

### 4.6.2 The surface tension of surfactant solutions

The *surface tension* of a surfactant solution, when plotted against concentration, usually falls smoothly to a lower limit after which it remains constant (see Figure 4.12). The curve is nearly linear at very low concentrations and the Gibbs equation (4.8) indicates that the adsorption is then proportional to the concentration. When the surface tension is plotted against $\ln c$ there is a linear fall just before an abrupt change to the constant value of surface tension. The Gibbs equation shows that where the $\gamma$–$\ln c$ curve is linear the absorption is constant, suggesting that the adsorption has reached a limit. This is often called the **saturation adsorption**. However, recent neutron reflectivity and surface tension data have brought the concept of saturation adsorption into question (Simister *et al.*, 1992; Barnes *et al.*, 1998).

### 4.6.3 Adsorption of surfactants

Adsorption values can be calculated from such surface tension data using the Gibbs equation (3.32 or 3.33). Usually it is the logarithmic form of the equation that is used, as calculation of the slopes is easier. The adsorption isotherm for $C_{16}TAB$ calculated from the data in Figure 4.12 is shown in Figure 4.13.

There are two approximations that may be useful for calculating adsorption when the data are sparse. At low surfactant concentrations (indicated by values of surface tension only slightly lower than that of water) the slope $d\gamma/dc$ is nearly constant, and at concentrations just below the region of constant surface tension the slope $d\gamma/d\ln c$ is nearly constant.

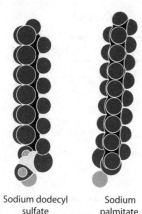

Sodium dodecyl sulfate    Sodium palmitate

Cetyl trimethyl ammonium bromide

**Figure 4.10.** Space-filling molecular models of several amphiphiles.

**Table 4.2.** Systematic and common names of some fatty acids and alcohols. Note that that some of these amphiphiles are not sufficiently soluble in water to be classed as surfactants.

| No. of carbons | Alcohol | | | Acid | | |
|---|---|---|---|---|---|---|
| | Systematic name | Common name | Abbreviation | Systematic name | Common name | Abbreviation |
| 12 | dodecan-1-ol | lauryl alcohol | $C_{12}OH$ | dodecanoic acid | lauric acid | $C_{11}COOH$ |
| 14 | tetradecan-1-ol | myristyl alcohol | $C_{14}OH$ | tetradecanoic acid | myristic acid | $C_{13}COOH$ |
| 16 | hexadecan-1-ol | cetyl alcohol | $C_{16}OH$ | hexadecanoic acid | palmitic acid | $C_{15}COOH$ |
| 18 | octadecan-1-ol | stearyl alcohol | $C_{18}OH$ | octadecanoic acid | stearic acid | $C_{17}COOH$ |
| 20 | eicosan-1-ol | arachic alcohol | $C_{20}OH$ | eicosanoic acid | arachidic acid | $C_{19}COOH$ |
| 22 | docosan-1-ol | | $C_{22}OH$ | docosanoic acid | behenic acid | $C_{21}COOH$ |

**Table 4.3.** Names of common unsaturated and ionic surfactants.

| Systematic name | Common name | Abbreviation |
|---|---|---|
| 9-octadecenoic acid (*cis*) | Oleic acid | |
| 9-octadecenoic acid (*trans*) | Elaidic acid | |
| 9,12-octadecadienoic acid | Linoleic acid | |
| Sodium dodecyl sulfate | Sodium lauryl sulfate | SDS |
| Hexadecyl trimethyl-ammonium bromide | Cetyl trimethyl ammonium bromide | $C_{16}TAB$ |
| Dodecyl penta(ethylene oxide) | | $C_{12}E_5$ |

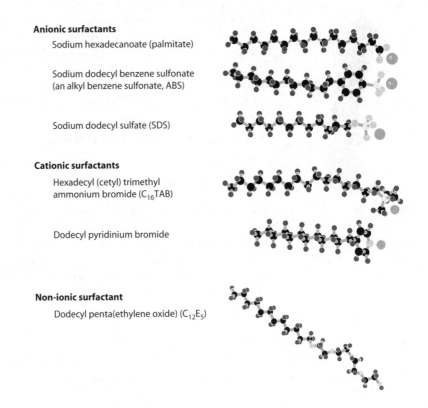

**Anionic surfactants**
　Sodium hexadecanoate (palmitate)

　Sodium dodecyl benzene sulfonate
　(an alkyl benzene sulfonate, ABS)

　Sodium dodecyl sulfate (SDS)

**Cationic surfactants**
　Hexadecyl (cetyl) trimethyl
　ammonium bromide ($C_{16}TAB$)

　Dodecyl pyridinium bromide

**Non-ionic surfactant**
　Dodecyl penta(ethylene oxide) ($C_{12}E_5$)

**Figure 4.11.** Examples of surfactants.

## Application of the Gibbs equation

Comparisons of adsorption values determined by the Gibbs equation and values determined by more direct methods occasionally reveal discrepancies.

Surface hydrolysis is clearly illustrated by the following example (Sally *et al.*, 1950). Adsorption measured by radio labelling of the anionic surfactant 'Aerosal OTN' (sodium di-*n*-octyl sulfosuccinate) with $^{35}S$ showed that Eq. (4.8) should be used without the factor $\frac{1}{2}$. This suggested that $Na^+$ was not being adsorbed and further measurements with $^{22}Na^+$ confirmed this. Apparently the $Na^+$ was being replaced at the surface by $H^+$, a phenomenon

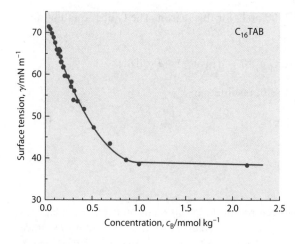

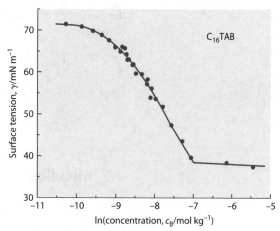

**Figure 4.12.** Surface tension of $C_{16}TAB$ solutions plotted against concentration and ln(concentration). (Data of G. A. Lawrie.)

now called *surface hydrolysis*. Applying the full Gibbs equation (3.30) (with $Na^+ = M^+$) gives:

$$-\frac{d\gamma}{RT} = \Gamma_{M^+} d\ln(c_{M^+}) + \Gamma_{X^-} d\ln(c_{X^-}) + \Gamma_{H^+} d\ln(c_{H^+}).$$

But $\Gamma_{M^+} = 0$ and $c_{H^+}$ is constant (the amount of $H^+$ lost to the surface is negligible compared to the total amount in the bulk solution), so the equation becomes

$$-\frac{d\gamma}{RT} = \Gamma_{X^-} d\ln(c_{X^-}). \tag{4.9}$$

### Indifferent electrolyte

Sometimes the solution also contains a relatively large concentration of an indifferent electrolyte with the same counterion as the amphiphile (e.g. $M^+Y^-$). This avoids the ambiguities that may arise with surface hydrolysis and often

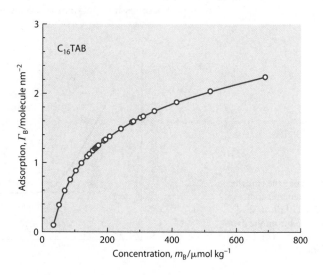

**Figure 4.13.** Adsorption isotherm for $C_{16}TAB$.

the electrolyte is added deliberately for this reason. The Gibbs equation then becomes:

$$-\frac{d\gamma}{RT} = \Gamma_{M^+} d\ln(c_{M^+}) + \Gamma_{X^-} d\ln(c_{X^-}) + \Gamma_{Y^-} d\ln(c_{Y^-}).$$

But $\Gamma_{Y^-} = 0$ and $d\ln(c_{M^+}) \approx 0$, resulting in:

$$-\frac{d\gamma}{RT} = \Gamma_{X^-} d\ln(c_{X^-}). \tag{4.10}$$

## 4.7 Micelles

Above a certain critical concentration the surface tension becomes independent of concentration, leading to a clear change in slope of the plot of surface tension against concentration (Figure 4.12), and is particularly marked in the logarithmic plot. This and other evidence indicates the formation of large aggregates known as **micelles**. The critical concentration is called the **critical micelle concentration** or **cmc**, and can be easily determined by the intersection of extrapolations of the linear regions each side of the change in slope (Figure 4.14). In an aqueous solution the amphiphile molecules in the micelles are arranged with their hydrophilic groups outwards and their hydrophobic segments inside, thus satisfying the amphiphilic requirements of the surfactant. In this structure the surface of the micelles is hydrophilic and consequently the micelles do not adsorb at the air–solution interface.

If $X_s$ represents single (non-ionic) surfactant molecules in solution and $X_m$ the micelles, the Gibbs equation becomes:

$$-\frac{d\gamma}{RT} = \Gamma_{X_s} d\ln(c_{X_s}) + \Gamma_{X_m} d\ln(c_{X_m}). \tag{4.11}$$

But $\Gamma_{Xm} = 0$ and $c_{X_s}$ remains nearly constant above the cmc because all of the excess amphiphile goes to form micelles (implying that $\ln c_{X_s}$ is constant and therefore $d\ln c_{X_s} = 0$). Thus both terms in Eq. (4.11) vanish, and the surface tension becomes independent of concentration.

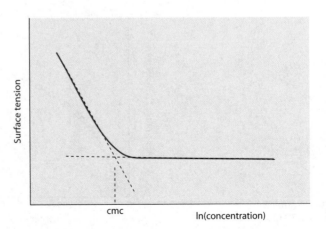

**Figure 4.14.** Above the cmc the surface tension remains constant. The intersection of the dashed lines extrapolated from the data allow a determination of the cmc from experimental data.

### 4.7.1 Molecular requirements

It is important to note that not all amphiphiles form micelles. The simple long-chain amphiphiles with a small uncharged head group (such as hexa-decanol) do not form micelles whereas similar molecules with charged head groups do. Molecules with large non-ionic head groups, such as the poly(oxyethylene)s, do form micelles. We note that these observations are in accord with concepts based on the *packing parameter* or *shape factor* which are discussed in more detail later (Section 6.7.3).

### 4.7.2 Formation of micelles

At the cmc, other properties of surfactant solutions also show distinct changes in behaviour. Some of these properties are shown in Figure 4.15.

The value of the cmc often shows some variation with the method of detection and even with the same method but different operators. Some of this variation can be attributed to impurities in the surfactant (surfactants are notoriously hard to purify) and some to the procedures used to calculate the cmc when the trend in the property changes over a range of concentration values. A comprehensive tabulation of values up to 1966 is given by Mukerjee and Mysels (1971) and some more recent values by Rosen (1978). Values for some of the more common surfactants are listed in Table 4.4.

Perhaps the main clue leading to an explanation of the behaviour patterns shown in Figure 4.15 is provided by the osmotic pressure. This is a colligative property and as such it depends not on the total concentration of solute but on the concentration of solute particles. Thus up to the cmc addition of solute yields one particle per added molecule (assuming that the solute is non-ionic), but beyond the cmc a large number of molecules is required to form one new particle. Clearly these newer particles are aggregates of molecules and the term *micelles* has been applied to them. This proposal also explains the patterns observed with the other properties in Figure 4.15. Some of these

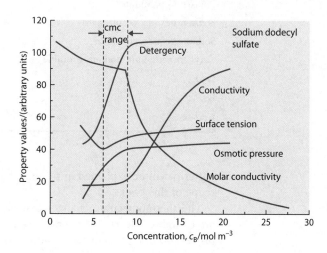

**Figure 4.15.** Various properties of a solution of sodium dodecyl sulfate plotted against surfactant concentration. Note the minimum in the surface tension curve indicating an impurity in the surfactant (see Figure 4.20). (Replotted from data of Preston, 1948.)

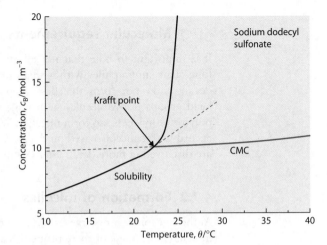

**Figure 4.16.** The temperature dependence of cmc and solubility showing the origin of the Krafft point. (Adapted from Shinoda *et al.*, 1963.)

**Table 4.4.** Critical micelle concentrations at $25\,^\circ$C (unless otherwise stated) and Krafft temperatures of some common surfactants. These data have come from a variety of sources and appreciable differences have been found where comparisons were possible.

| Surfactant | Cmc/mol m$^{-3}$ | Krafft temperature, $T/^\circ$C |
|---|---|---|
| Sodium dodecanoate | 26.0 | 19 |
| Potassium dodecanoate | 25.5 | |
| Sodium dodecyl sulfate | 8.1 | 9 |
| Sodium dodecyl sulfonate | 9.8 | 22 |
| Potassium dodecyl sulfonate | 9.0 | |
| Sodium palmitate | 3.2 (at $50\,^\circ$C) | 48 |
| Potassium palmitate | 2.2 (at $50\,^\circ$C) | |
| Sodium hexadecyl sulfonate | 0.75 (at $50\,^\circ$C) | |
| Sodium hexadecyl sulfate | 0.58 (at $40\,^\circ$C) | 34 |
| Cetyl trimethyl ammonium bromide | 0.96 | |
| $C_{12}E_4$ | 0.04 | |
| $C_{12}E_7$ | 0.05 | |
| $C_{12}E_{23}$ | 0.06 | |

patterns will be explained in subsequent sections, but others are beyond the scope of this book.

The thermodynamic changes associated with the formation of micelles are complex. Clearly the overall free energy change must be negative as the process occurs spontaneously, but it seems that the enthalpy change is small, and sometimes positive, whereas (perhaps surprisingly) there is a large increase in entropy. At first sight, one would expect a decrease in entropy as free surfactant molecules aggregate into a structured micelle, but there is a much larger increase in the entropy of the aqueous solvent. In water, the alkyl chain of a surfactant molecule induces a water structure that surrounds the chain, reminiscent perhaps, of a clathrate, and known as the *hydrophobic effect* (see Section 2.9.3). It is the large entropy increase when this structure breaks down during micelle formation, offset to a small extent by the re-establishment of the normal hydrogen-bonded structure of liquid water, that determines the large entropy increase.

### Solubility of surfactants

At low temperatures the solubility of an ionic surfactant is low, but when the temperature is raised to a characteristic temperature, known as the **Krafft temperature**, there is a sharp rise in solubility. This is the lowest temperature at which the solubility exceeds the cmc. Figure 4.16 shows the temperature dependence of both the cmc and the solubility and explains the origin of the Krafft point.

The Krafft temperatures of a few common surfactants are given in Table 4.4.

With non-ionic surfactants, on the other hand, raising the temperature appears to decrease the solubility so that at a certain temperature, known as the *cloud point*, large aggregates of the surfactant separate out as a distinct phase and the liquid becomes cloudy. These systems thus exhibit a *lower consolute temperature*.

### 4.7.3 Micelle structure

In dilute solutions the micelles are spherical, with a diameter slightly less than twice the length of the amphiphile molecule and containing between 20 and 50 molecules. As noted above, the amphiphile molecules are arranged in micelles with their hydrophilic groups towards the water and their hydrophobic tails in the micelle interior away from the water. With ionic surfactants the counterions are important in reducing repulsion between the charged head groups in the micelle surface. A computer simulation of a spherical micelle is shown in Figure 4.17.

At higher concentrations cylindrical micelles form and then more complex structures (see Figure 4.18).

In dealing with surfactants the solvent is generally water, but occasionally organic solvents are used. In those cases the micelle structure is inverted, with the oleophilic (hydrophobic) groups on the outside and the hydrophilic groups inside. These are sometimes called reverse or inverted micelles, and are shown on the right-hand side of Figure 4.18.

### 4.7.4 Solubilization

For spherical micelles in particular, the centre of the micelle is essentially a liquid hydrocarbon, and as such it is capable of dissolving some oil-soluble

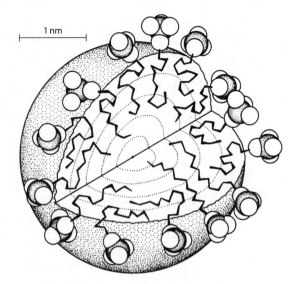

1 nm

**Figure 4.17.** Computer simulation of a micelle of sodium dodecylsulfate containing 60 molecules. The ionized head groups are outside the hydrophobic core of the micelle, and the hydrocarbon chains are in a liquid-like state. (From Israelachvili 1991.)

substances. Thus solutions containing micelles sometimes increase the solubility of sparingly soluble substances by incorporating these molecules into the micelles, a process called **solubilization**. Hydrophobic solutes tend to be solubilized in the hydrophobic centres of the micelles while amphiphilic solutes may be incorporated into the micelle structure in much the same way as the surfactant molecules themselves.

Example

Rodriguez and Offen (1977) measured the solubility of naphthalene in sodium dodecyl sulfate solutions. Their results show a sharp rise at the cmc (Figure 4.19).

The apparent cmc obtained from the solubilization was $3\,mol\,m^{-3}$, a value that is appreciably lower than the values obtained by other methods: about $8\,mol\,m^{-3}$ (Table 4.4). This is attributed to the presence of naphthalene which acts like an impurity in the surfactant and lowers the cmc (see Section 4.7.6).

### 4.7.5 Factors affecting the cmc

The cmc is affected by factors that alter the solubility of single molecules and by factors that change the ease of micelle formation: decreasing solubility decreases the cmc while structures that hinder packing and repulsion between like ionic charges increase the cmc.

- *Amphiphile chain length* ($\lambda$). For a homologous series, the longer the alkyl chain the less soluble are the molecules in water so that micelles tend to form at lower concentrations:

$$\ln\,(\text{cmc}) = a - b\lambda$$

where $a$ and $b$ are constants.

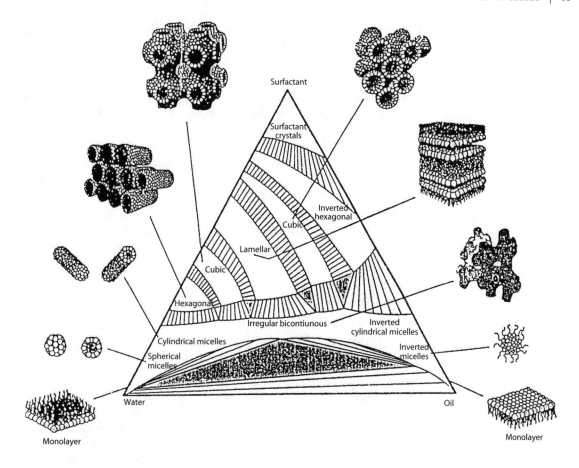

**Figure 4.18.** Schematic diagrams of the microstructures associated with a schematic oil–water–surfactant phase diagram. (From Vinson *et al.*, 1991.)

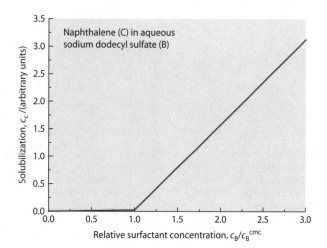

**Figure 4.19.** Solubilization of naphthalene in sodium dodecyl sulfate solutions. (Replotted from data of Rodriguez and Offen, 1977.)

- *Salts* have a significant effect on the cmc of ionic surfactants. The presence of a dissolved salt lowers the cmc due primarily to screening of the head-group charges in the micelle surface, thus decreasing the repulsion between like charged head groups.
- *Head group structure.* Size and shape of the head group does not change the cmc very much as there is normally ample room at the micelle surface, but for straight-chain ionic surfactants the cmc does vary through a range with a factor of about two.
- *Structure of the alkyl chain.* An increase in the bulkiness of the hydrophobic moiety generally raises the cmc, but a benzene ring in the alkyl chain lowers it.
- *Other surfactants.* If the two cmc values are similar the mixtures all have the same cmc. If they differ greatly, the surfactant with the higher cmc acts like an electrolyte.
- *Polar additives.* Long-chain materials with a polar group markedly lower the cmc and the additive is solubilized by being incorporated into the micelle structure.

### 4.7.6 Impurity effects

Some early data showed a minimum in the surface tension–concentration curve like that shown in Figure 4.15 and some of the curves in Figure 4.20.

These curves, except that for the purified surfactant, show positive slopes on the high concentration side of the minimum and according to the Gibbs equation this feature would be interpreted as a negative adsorption. However the surface tension at these concentrations is low indicating that there is an appreciable concentration of surfactant at the surface. Careful purification of the surfactant removed the minimum, showing that it was an impurity in the surfactant that was responsible. It has since been shown that sodium dodecyl

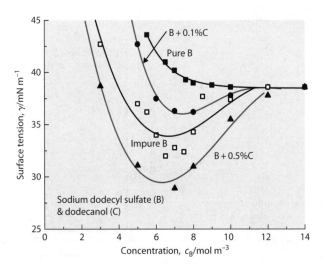

**Figure 4.20.** Surface tension data for purified sodium dodecyl sulfate with various amounts of dodecanol (a likely impurity) and for unpurified surfactant. (Replotted data of Miles and Schedlovsky, 1944.)

sulfate (SDS) is easily hydrolysed giving dodecanol as one of the products. Thus this amphiphile is the impurity responsible for the minimum, as shown in Figure 4.20.

The explanation of the minimum is as follows. The impure SDS lowers the surface tension to a slightly greater extent than pure SDS until micelles start to form. (A similar effect is seen in the penetration of insoluble monolayers by soluble surfactants – Section 5.7.2.) As the micelles form they incorporate the dodecanol impurity by solubilization and effectively remove it from solution as the micelle concentration rises. This causes the surface tension to rise back to the value it would have with pure SDS.

### Tests of surfactant purity

The absence of a minimum is sometimes taken as an indication of the purity of a surfactant. A more sensitive test has been proposed by Lunkenheimer and Miller (1979). After a solution of the surfactant has established adsorption equilibrium, the area of the surface is changed suddenly and the consequent change in surface tension monitored until a steady value is reached. The area is then changed in the opposite direction and the surface tension again followed. With a pure system the surface tension rapidly returns to the equilibrium value each time, but when impurities are present the rate of return is slow and often incomplete.

## 4.8 Applications of surfactants

The manufacture of surfactants is now a major commercial enterprise. For example, in North America surfactant consumption was estimated to be about $2 \times 10^9$ kg in 2002 and rising (Anon, 2003). In this section, some of the major applications of surfactants are briefly reviewed. In the following section, the very important applications of films and foams are discussed in detail.

### 4.8.1 Wetting

The wetting of a solid surface by a liquid was discussed earlier in Section 2.3, where it was shown that the contact angle and the ability of the liquid to spread over the solid surface are determined by the relevant interfacial tensions. It is clear that lowering the surface tension of the liquid and the interfacial tension between liquid and solid would lower the contact angle and increase the tendency of the liquid to spread over the solid.

Many surfactants are able to produce such changes in the interfacial tensions and consequently they find applications in paints, agricultural sprays, animal dips, printing inks, dyestuffs, lubricants, detergents, and so on.

Umbrella cloth is now made from nylon which is hydrophobic. Even though there are gaps between the fibres of the cloth it nevertheless keeps the rain out because of the high contact angle. The effect can be demonstrated with the apparatus shown in Figure 4.21.

**Figure 4.21.** Apparatus for demonstrating the effect of a detergent on the wetting of umbrella cloth.

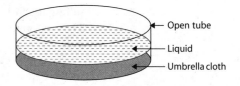

**Figure 4.22.** Schematic microscopic view of water and detergent solution on nylon cloth (blue). There is a large contact angle with water and a small contact angle with detergent solution enabling the liquid to move through the cloth.

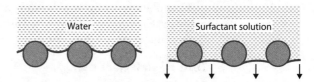

If pure water is put into the tube it does not penetrate through the umbrella cloth, but if a surfactant solution is used it comes through and drips out. The principle is shown schematically in Figure 4.22.

### 4.8.2 Detergency

The term *detergent* will here be reserved for mixtures of surfactants with other materials in order to make an effective cleansing formulation. Cleaning involves the removal of *dirt* or *soil* which can be defined as matter in the wrong place. For example, tomato sauce on your food is good, but on your clothes it becomes dirt and should be removed. Dirt is usually a very complex mixture: it mostly contains some particulate matter and possibly oils, fats, greases, and proteins. Dirt that is hydrophilic is readily dispersed by rinsing with water, so the subject of detergency, and our current discussion, is largely concerned with hydrophobic dirt. We will use the term *substrate* for the surface to be cleaned. It may be a hard solid surface like a dinner plate or a soft surface like a fabric. Detergency nearly always involves agitation as the shearing action helps to separate the particles of dirt from the substrate.

The basis of any detergent is the *surfactant*, which may be a soap (such as sodium palmitate) or a synthetic material (such as those in Figure 4.11). The surfactant has a number of functions in a detergent formulation. First of all it promotes the wetting of the substrate by adsorbing at the interface between the substrate and the detergent solution. Secondly it also adsorbs on to the dirt tending to emulsify it. Thus both the substrate and the soil become coated with an adsorbed layer of detergent and this promotes the penetration of the detergent solution between the substrate and the soil. This will occur spontaneously if the work of adhesion is negative (see Eq. 2.7):

$$w_{so} = -\gamma_{so} + \gamma_{sd} + \gamma_{od} \tag{4.12}$$

where s, o, and d refer to substrate, oil (soil), and detergent solution respectively. Thus the reduction of $\gamma_{sd}$ and $\gamma_{od}$ by the adsorption of surfactant decreases $w_{so}$ and aids soil removal. It is worth noting that the surface tension of the air–detergent solution interface does not appear in Eq. (4.12) so the ability of a surfactant to produce a foam does not necessarily indicate that it will be an efficient cleaning agent.

Micrographs of detergent action show that an oily dirt is rolled up by the detergent and eventually separated from the substrate as emulsified particles. The coating of surfactant on the substrate also helps to prevent the emulsified dirt from redepositing on the substrate.

Reference to Figure 4.15 shows that the detergent activity of a surfactant rises steeply with concentration up to the cmc after which it levels out. This indicates that, despite the possibility of solubilization of soil, the micelles do not play a direct role in detergent action. Their chief function is probably to provide a reservoir of single surfactant ions or molecules to replace those removed by participation in the cleansing process.

A commercial laundry detergent contains:

- a surfactant system;
- builders;
- brighteners and bleaches;
- electrolyte filler;
- sales appeal ingredients.

The **surfactant** system is usually a mixture of several surfactants as it is found that there is synergetic effect in such mixtures. A common surfactant in domestic laundry detergents is an anionic, sodium alkyl benzene sulfonate, mixed with a non-ionic. The **builders** aid the action of the surfactants. Usually a major builder is sodium tripolyphosphate which buffers the wash liquid, acts as a sequestering agent for $Ca^{2+}$ and $Mg^{2+}$ which interfere with the action of even synthetic surfactants, and aids in the suspension of clay type soils. Another builder is sodium silicate which helps control alkalinity, inhibits corrosion, and aids in the manufacturing process. Yet another builder is sodium carboxy-methyl cellulose which is an anti-redeposition agent.

The **bleach** is usually sodium perborate which gives off nascent oxygen when heated in water. This has a bleaching action. **Brighteners** are usually fluorescent materials that absorb ultraviolet light and re-emit it at the blue end of the visible spectrum. A blue tinge is commonly associated with cleanliness. The **electrolyte filler** in domestic laundry powders is usually sodium sulfate which lowers the cmc of the surfactants by increasing the ionic strength (see Section 4.7.5) and thereby increases the activity of the surfactants so that less is needed. It also helps prevent the powder from caking. Ingredients that add colour and perfume may also add sales appeal but do little to enhance the detergent action.

A **fabric softener** may also be used, often in the rinse stage. This is usually a cationic surfactant which adsorbs strongly on the negatively charged fabric with its alkyl chains outward. This coating provides the lubrication between fibres that is sensed as a softening of the fabric. The carboxylic acid soaps have a similar effect, but may also incorporate calcium ions which build up as a film of dirt.

The substantial replacement of laundry soaps by synthetic surfactants in the 1950s and 1960s initially caused major pollution problems, but these have now been substantially solved. The biodegradability of surfactants has been improved by replacing branched alkyl chains by linear chains and phosphate pollution (which caused algal blooms) has been reduced by substituting materials such as nitrilo triacetic acid, zeolite A, or poly-carboxylates.

### 4.8.3 Water repulsion

A hydrophilic solid surface can be rendered hydrophobic by adsorbing a suitable surfactant onto it. The surfactant will generally adsorb with its hydrophilic head group on the surface and the hydrophobic tails projecting away from the surface. Thus if sufficient surfactant has been adsorbed, the surface is covered by the hydrophobic tails and becomes hydrophobic. $C_{16}TAB$, for example, will adsorb strongly onto a clean glass surface and because it does not desorb readily, the glass becomes hydrophobic. In some such cases the solid is no longer wet by the parent liquid and the film is then described as autophobic.

### 4.8.4 Emulsification

An emulsion is a dispersion of one liquid phase in another liquid. Such dispersions are usually unstable, tending to separate into two layers of liquid, unless a stabilizing agent is added. Most stabilizing agents, usually called emulsifiers, are surfactants. This application will be discussed in Section 6.3.

### 4.8.5 Froth flotation in ore treatment

The purification of mineral ores by froth flotation is an important industrial process and huge amounts are processed annually. The impure mineral is powdered so that the particles consist of either mineral or waste material and suspended in an aqueous solution through which air bubbles are blown. Particles with a hydrophobic surface are able to stick to the bubbles and are thus carried upwards into the foam that is formed. The solution contains a collector designed to ensure that the mineral particles are hydrophobic while leaving the contaminant particles (known as the *gangue*) hydrophilic and a frother which enables the formation of a foam with sufficient stability for skimming off, with the mineral, to a receptacle. Activators and depressants may also be present to modify the action of the collector. The basic science has been described by Wark (1979).

The selection of a collector and frother appropriate for the ore to be separated is extremely important and while a lot of selection work has been done empirically a detailed study of the chemistry of the ore is highly desirable.

Much of the early work concerned lead and copper sulfide ores and for these the xanthates were found to be very effective as collectors. There is, however, some evidence that the effective entity is dixanthogen:

$$R-O-(C{=}S)-S^- \quad R-O-(C{=}S)-S-S-(C{=}S)-O-R$$
Xanthate \qquad Dixanthogen

For oxide ores the surface charge is controlled by the pH of the solution and simple ionized surfactants will adsorb at the appropriate pH: cationic at high pH, and anionic at low pH values. Clearly the ionic group of the surfactant is attached to the mineral with the hydrophobic segments outwards making the mineral particles hydrophobic and floatable.

Some solids are naturally hydrophobic and therefore floatable. If flotation is not desired it may be depressed by chemically altering the surface: extensive oxidation or adsorption of hydrophilic colloids, such as starch, may be used to render the solid surface hydrophilic.

Coal is naturally hydrophobic and therefore can be floated, while silicates, a common impurity are naturally hydrophilic and are not floated. Sulfur is a common, but unwanted impurity in coal and its removal is highly desirable. Organic sulfur occurs within the coal structure and cannot be separated by flotation, but inorganic sulfur (mainly pyrite and marcasite) can be separated. One strategy involves two stages: a normal flotation to remove the silicates and similar gangue materials, followed by sulfur removal using a depressant to prevent coal flotation and a sulfhydryl collector to cause the pyrite to float.

### 4.8.6 Oil recovery

When oil is removed from an oil-bearing rock a significant amount is left behind. This is a valuable resource and the possibility of its recovery has generated a considerable research effort. The basic principle is similar to that in detergency (see Eq. 4.12): pumping a surfactant solution into the oil-bearing rock dislodges the oil from the rock and disperses it in the solution so that it can be pumped out as a mixture of oil and water.

### 4.8.7 Membrane disruption

There are many applications of surfactants in biology and some will be described later in Chapter 10. Only an illustrative example will be given here.

Some surfactants appear capable of disrupting natural membranes. For example, several surfactants promote the penetration of hexylresorcinol into the pig round worm, but as the concentration of surfactant is raised the effect reaches a maximum and then declines. This pattern is attributed to the formation of micelles and is explained in detail in Section 10.5.4.

## 4.9 Films and foams

### 4.9.1 Introduction

The ability of surfactants to form soap films is well known. Soap films have been a source of delight to everyone from small children to mature and serious scientists and mathematicians. For children it is usually the forms and inter- ference colours that fascinate; for scientists it is the challenge of explaining film formation and behaviour; for mathematicians it is the ability of such films to form minimum area surfaces that provides inspiration.

Soap films consist of a thin layer of solution with an adsorbed layer of surfactant at each face (Figure 4.24). The surfactant molecules are, of course,

---

**A simple example**

A simple example of the use of soap films by mathematicians has been given by Isenberg (1978). The problem is to join four towns, situated at the corners of a square, by a road of minimum length. The analogue solution of the problem is obtained by using four pins to represent the towns and sandwiching them between two perspex plates. A soap film between the plates and joining the four pins is formed by immersing the structure in a soap solution and allowing the film to come to equilibrium. A structure like that shown in Figure 4.23 is formed and represents the solution to the problem. The reader can visualize other arrangements and compare the different lengths of road.

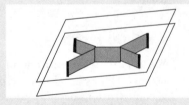

**Figure 4.23.** Minimum length of a soap film stretched between the four corners of a square. At the two places where three films join the angles between the films are all 120°.

---

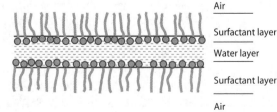

Air

Surfactant layer

Water layer

Surfactant layer

Air

**Figure 4.24.** Schematic diagram showing the structure of a soap film.

oriented with their hydrophilic heads towards the aqueous centre of the film and their hydrophobic tails towards the air.

If a soap film is held vertically, the central liquid tends to drain out and the film thins until the process is halted by the repulsion between the two adsorbed surfactant films. During the draining process the film thins from the top and optical interference produces spectacular colours which change and move downward. Eventually the film becomes silver and then black as the optical interference becomes destructive for all visible wavelengths. Two equilibrium black films can be distinguished: **common black films (CBF)** have a thin layer of liquid between the two surfactant films, whereas **Newton black films (NBF)** have no free liquid core. Transitions in both directions between these two types of black film are possible. Where the film is held by its supporting frame the film is thicker and during drainage there are often spectacular and dynamic colour effects where the main film contacts the thicker margin.

### 4.9.2 Film tension

A soap film has a film tension and consequently is subject to the requirements of the Laplace equation (2.15). This can be demonstrated by the experiment described in Figure 2.19, where two soap bubbles of different size are connected by a tube. The larger bubble expands and the smaller contracts to form a film across the end of the tube with the same radius of curvature as that of the large bubble.

The magnitude of the film tension can be measured by using the Laplace equation. A known pressure difference is applied across the film and the resulting curvature measured.

When there is an appreciable thickness of bulk liquid between the two adsorbed surfactant layers the film tension is simply twice the tension of the adsorbed film at the surface of the bulk liquid phase. However, when the film thins and a black film forms there is significant interaction between the two adsorbed surfactant layers and this can change the overall film tension. The subject is too complex to treat fully here.

If a film is stretched there will tend to be a general increase in film tension, but momentarily at least this may not be uniform. If the film stretches more at one place it will be thinner there and the surface tension higher than elsewhere. As a consequence, the gradient in surface tension tends to draw surfactant film towards the thinner spot and this movement drags bulk solution with it. Thus additional surfactant is brought in to replenish the surface and additional bulk fluid restores the thickness of the film. Clearly the presence of surfactant is essential to the stability of the film.

### 4.9.3 Permeability to gases

The ability of gases to pass through a soap film is an important factor in the stability of a foam. It is described by a permeability coefficient, $k_f$, or a permeation resistance, $r$ $(= 1/k_f)$. Three stages in film permeation can be distinguished: passage through the first adsorbed surfactant layer, passage through the central solution layer, and passage through the second adsorbed surfactant layer.

Monolayer permeation is discussed in more detail in Chapter 5 and for the present discussion can be characterized by a permeation constant, $k_m$:

$$J = k_m \Delta c_m \tag{4.13}$$

where $J$ is the flux, the rate at which gas passes through unit area, and $\Delta c_m$ is the concentration difference across the adsorbed monolayer.

Diffusion through the bulk solution layer is described by Fick's first law of diffusion:

$$J = k_b \Delta c_b = (DH/z_b) \Delta c_b \tag{4.14}$$

where $D$ is the diffusion coefficient, $H \, (= c^L/c^G)$ is Henry's law factor required to convert gas concentrations in the liquid phase to their equilibrium concentration in the gas phase, $z_b$ is the thickness of the bulk liquid layer, and $\Delta c_b$ is the concentration difference across the bulk liquid layer expressed as gas phase concentrations.

Rearranging and then combining Eqs. (4.13) (twice) and (4.14) gives for steady state permeation:

$$J\left(\frac{1}{k_m} + \frac{z_b}{DH} + \frac{1}{k_m}\right) = \Delta c \tag{4.15}$$

or

$$J = k_f \Delta c \tag{4.16}$$

where $\Delta c$ is the difference in gas concentration across the entire film and $k_f$ is the overall permeation coefficient.

The experimental measurement of film permeation has usually been performed by observing the shrinkage of a gas bubble trapped at the surface of a surfactant solution (Princen *et al.*, 1965, 1967).

### 4.9.4 Foam formation

A foam is an assembly of soap bubbles with adjacent bubbles sharing the common soap film. Often a foam is formed by passing gas bubbles into a surfactant solution. In the bulk of the solution there is adsorption of surfactant at the bubble–solution interface. These bubbles rise to the surface which is also covered by an adsorbed layer of surfactant so that a bilayer of surfactant is formed (Figure 4.25). Successive bubbles rising to the surface contact a second surfactant layer from either the gas–solution interface or previously formed bubbles. Eventually, if the bubbles are sufficiently stable, the surface is covered by bubbles and a foam is formed. By analogy with other disperse systems this method can be classified as a dispersion method.

Condensation or aggregation methods are also used. The bubbles formed in carbonated drinks or sparkling wines when the pressure is released are common examples.

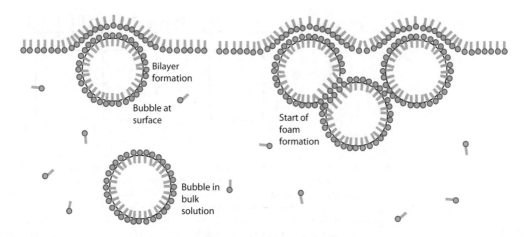

**Figure 4.25.** Processes in foam formation.

Different surfactants will produce different amounts of foam under identical conditions. Usually the relative foaming abilities of surfactant solutions are measured by the height of foam produced in a vertical tube. Obviously such measurements are affected by the stability of the foam. Foam volume is one component of foam stability, but other properties, such as bubble size and liquid content are also relevant. A simple comparative measurement of foam stability is the foam lifetime.

### 4.9.5 The Plateau border

At the junctions between soap films in a foam there is a region of greater thickness known as a Plateau border (Figure 4.26).

The Plateau borders in a foam make an important contribution to foam drainage. Application of the Laplace equation shows that because of the different curvatures the pressure in the Plateau border is lower than that in the foam lamellae. Consequently liquid will tend to flow from the lamellae into the Plateau borders promoting film drainage, thinning the soap films, and reducing the stability of the foam. If this process continues without the films rupturing a stage is eventually reached where the two adsorbed layers of soap are very close. A repulsive pressure preventing closer approach may be experienced at this stage, known as the disjoining pressure.

Recent treatises on surfactant films and foams have been produced by Ivanov (1988) and by Exerowa and Kruglyakov (1998).

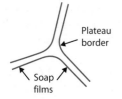

**Figure 4.26.** Junction of soap films showing the Plateau border. Usually in the transition from Plateau border to thin film the surface is not a continuous smooth curve but exhibits a small contact angle.

## 4.10 The Marangoni effect

The observation of 'tears of wine' was a popular pastime in ancient Rome but was not explained until 1855 (Thompson, 1855). If the inside wall of a wine glass is wetted by the liquid, the alcohol evaporates more rapidly than the water and as this occurs unevenly those regions of more rapid evaporation

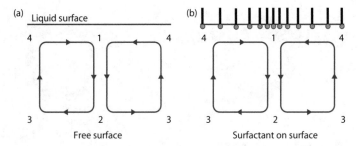

**Figure 4.27.** Schematic diagram of convection cells. Due to evaporation, the liquid flows in the directions indicated. For buoyancy convection, $T_1 < T_2$ so $\rho_1 > \rho_2$ and $T_3 > T_4$ so $\rho_3 < \rho_4$; for Marangoni convection $T_4 > T_1$ so $\gamma_4 < \gamma_1$. With a free surface (a) these flows are impeded by the viscosity, but if surfactant (or a surface contaminant) is present (b) the convection draws surfactant to position 1, lowering the surface tension there and weakening the Marangoni convection. (Adapted from Barnes, 1978.)

become richer in water and hence have a higher surface tension than the other regions (see Figure 4.3). The higher surface tension draws liquid towards these regions leading to thicker tear-shaped droplets of wine around the wine glass just above the liquid surface. This and similar effects caused by differences in surface tension arising from differences in concentration were studied in the later part of the nineteenth century.

Marangoni (1871, 1878) published a theory of liquid flow arising from gradients in the surface tension, now known as **Marangoni convection**. Such gradients could originate from differences in composition and also from differences in temperature (see Figure 2.20).

In the evaporation of liquids, convection cells can arise in two ways. Cooling of the surface due to the latent heat of vaporization usually leads to an inverse density gradient so that eventually zones develop where cooler denser liquid plunges down from the surface and convection cells form. This is **buoyancy-driven convection** and occurs when the level of instability, calculated according to a theory developed by Rayleigh (1916), exceeds the forces resisting motion. The cooling of the surface can also set up temperature gradients in the surface and hence gradients in surface tension. These too can lead to the development of convection cells. The two mechanisms reinforce one another as can be seen in Figure 4.27a. Convection begins when the forces leading to convection are sufficient to overcome the viscous and inertial[3] forces that tend to damp out the movement.

The damping effect of surfactants or other contaminants on the surface can be substantial. Reference to Figure 5.3 indicates that the surface tension can decrease sharply for only a small increase in surface concentration so that the tendency for surfactant to be concentrated near position 1 (Figure 4.27(b)) may produce a strong trend for surface-tension-driven

3. Although inertial forces are often mentioned, the equations used to calculate the Rayleigh and Marangoni numbers that are used to determine whether convective flow will occur include the dynamic viscosity but do not include any term for inertia.

Marangoni flow in the direction opposite to that of temperature-driven Marangoni flow. Thus contamination of the surface (not necessarily with a surfactant) can lead to the weakening or even the absence of Marangoni flow.

Viewed from above, convection cells can have a variety of structures (Berg *et al.*, 1966a). A regular arrangement of hexagonal cells, known as Bénard (1901) cells, is often observed when the depth of the evaporating liquid is small (<3 mm), but gives way to other structures, such as rolls and streamers, at greater depths. However, water behaves differently as no convection has been reported for depths less than about 10 mm (Berg *et al.*, 1966b). This suggests that Marangoni convection does not occur in water, but, as always, it is difficult to rule out the possibility that the water surfaces were contaminated.

## 4.11 Aerosols

### 4.11.1 Occurrence

Aerosols are dispersions of liquid or solid particles in air. They are commonly experienced in nature as fogs and mists, which consist of liquid droplets, and as smoke, where the particles are solid. In the home, aerosol sprays are used for a variety of purposes, distributing perfumes, pesticides, paints, lubricants, ironing aids, medications, and so on. Agricultural sprays are an important means for distributing pesticides and weedicides over large areas. A low-cost method for removing salt from seawater involves the generation of an aerosol in warm air: the water evaporates from the droplets, the salt is deposited, and the purified water collected by cooling the air stream. Some industrial processes use aerosols, but often in industry the formation of aerosols is a nuisance requiring, for example, the precipitation of smoke particles before waste gases can be vented through a chimney. Aerosols are also produced by internal combustion engines, with the emission of solid particulate matter by diesel engines being a major health concern.

### 4.11.2 Formation

Aerosols can be prepared by dispersion methods or by aggregation methods.

The principal **dispersion method** is to break up a stream of liquid by a rapid stream of air. In domestic sprays, for example, the stream of air or propellant passes over a tube through which the liquid is sucked by Venturi action. In another method a fine stream of liquid has a high electrical potential applied to it. This breaks up the liquid stream into aerosol droplets.

In **aggregation methods** a supersaturated vapour is allowed to condense on nuclei. In the atmosphere, air containing water vapour may rise to higher altitudes, expand and cool possibly to a temperature sufficiently low for supersaturation to occur. The Kelvin equation (2.19) dictates the degree of supersaturation required for self or homogeneous nucleation, but at lower degrees of supersaturation dust particles in the atmosphere may act as condensation nuclei and clouds form as a result.

In such nucleation processes electrically charged nuclei are generally more effective than uncharged nuclei and even gaseous ions can act as nuclei. This is the principle of the cloud-chamber method for detecting ionizing radiation: the radiation passes through a chamber supersaturated with water vapour and condensation occurs on the ions formed by the radiation, showing the paths.

### 4.11.3 Monodisperse aerosols

The essential requirement in the production of monodisperse aerosols is to limit the concentration of nuclei. In the apparatus developed by Sinclair and LaMer (1949) a heated wire coated with sodium chloride provides nuclei to a stream of warm gas saturated with the aerosol material. The stream then cools as it passes up a long chimney and condensation occurs on the nuclei. Careful control of the temperatures at the various stages and of the nucleation process ensures that no self nucleation occurs and so droplets only grow on the nuclei and at a controlled rate. Consequently the droplets grow uniformly giving a very narrow range of sizes described as a monodisperse aerosol.

### 4.11.4 Electrostatic precipitation

For smoke in particular, the coagulation of the particles is important in reducing the contamination of the atmosphere for many industrial processes. One frequently used procedure is electrostatic precipitation. Air together with the aerosol particles is passed into an ionization chamber where it passes wires charged to a very high DC potential (12,000 to 15,000 V) which causes ionization of the air and adsorption of these ions on to the particles. The now highly charged particles are then passed through an array of plates that are alternately charged (2,000 to 5,000 V DC) and grounded where they are attracted to the earthed plates, discharged, and deposited.

### 4.11.5 Verification of the Kelvin equation

The Kelvin equation (2.19)

$$\ln\left(\frac{p^c}{p^\infty}\right) = \left(\frac{\gamma \overline{V}^L}{RT}\right)\left(\frac{2}{r_m}\right) \tag{2.19}$$

describes the dependence of the vapour pressure of a liquid on the curvature of the vapour–liquid interface. Testing this equation has been difficult. For instance, the curvature of liquid surfaces in capillaries is complicated by interaction with the capillary walls, but if the liquid is in the form of an aerosol such complications are avoided.

The procedure used by Sinclair and La Mer (1949) uses monodisperse aerosols and the measurement of particle size from scattering angles in the higher order Tyndall spectra (see Section 9.8.2). A monodisperse aerosol of a non-volatile liquid (such as dioctylphthalate) is prepared and its particle size measured. The aerosol is then introduced into a vessel that also contains a liquid mixture of the non-volatile liquid (dioctylphthalate) and a volatile liquid (such as toluene). The aerosol is allowed to come to equilibrium with the vapour of the solution. Initially some of the volatile liquid is transferred from the bulk liquid through the vapour phase to dissolve in the aerosol droplets. This continues until the vapour pressure of the volatile liquid in the droplets equals its pressure in the vapour phase which in turn is determined by the equilibrium vapour pressure of the bulk solution. During this process the aerosol droplets grow in size until equilibrium is reached. Measurement of droplet size then enables the concentration of volatile liquid in the droplets to be calculated. There is then sufficient data to test the Kelvin equation (see Exercise 4.7).

## SUMMARY

Equilibrium adsorption at the gas–liquid interface is usually calculated from surface tension measurements using the Gibbs equation although surface analysis is possible in some special cases. Measurement of adsorption kinetics by measuring the changes in surface tension at a freshly formed surface or by studying the behaviour after an equilibrated surface has been disturbed by a small transverse or longitudinal wave motion is described.

The adsorption of non-electrolytes roughly follows the Langmuir isotherm at very low concentrations, but at higher concentrations the competitive adsorption of the solvent must be considered. For simple inorganic ions in water there is often a rise in surface tension with concentration indicating negative adsorption.

Amphiphiles are molecules that possess both a hydrophilic group and a hydrophobic group. They are strongly adsorbed at the air–aqueous solution interface as the hydrophilic part of the molecule can lie in the water while the hydrophobic part is in the air. The surfactants are an important class of amphiphile distinguished by their ability to form aggregates in solution. These aggregates are known as micelles and they form spontaneously when the concentration of surfactant exceeds a critical value: the critical micelle concentration or cmc. Surfactants are the basic ingredient in detergents, they stabilize emulsions, and they are also able to form soap films and foams.

Aerosols, minute liquid (or solid) droplets dispersed in air, have applications in industry, agriculture, and domestic situations. They are also occur naturally as fogs and mists, clouds, and as smoke.

## FURTHER READING

Boys, C. V. (1959). *Soap Bubbles: Their Colours and the Forces Which Mould Them.* Dover Publications, New York. This reprinting of the revised edition of 1911 describes numerous experiments that were originally presented in lectures to school students. It is an entertaining and informative text that has become a classic.

Exerowa, D. and Kruglyakov, P. M. (1998). *Foam and Foam Films: Theory, Experiment, Application*. Elsevier, Amsterdam. An advanced and comprehensive treatment.

Fuerstenau, M. C., Miller, J. D., and Kuhn, M. C. (1985). *Chemistry of Flotation*. Society of Mining Engineers, New York.

Hamley, I. W. (2000). *Introduction to Soft Matter: Polymers, Colloids, Amphiphiles, and Liquid Crystals*. Wiley, Chichester. The chapter on amphiphiles provides a useful introduction to the subject and includes the necessary background surface science.

Isenberg, C. (1978). *The Science of Soap Films and Soap Bubbles*. Tieto Ltd., Clevedon, UK. The emphasis is on the use of soap films to help solve mathematical problems relating to minimal surfaces.

Ivanov, I. B. (ed.) (1988). *Thin Liquid Films: Fundamentals and Applications*. Marcel Dekker, New York. An edited assembly of chapters by nearly two dozen authors.

Jones, M. N. and Chapman, D. (1995). *Micelles, Monolayers, and Biomembranes*. Wiley-Liss, New York.

Rao, S. R. (2004). *Surface Chemistry of Froth Flotation, 2nd edn*. Kluwer Academic/Plenum. Describes the fundamental principles and latest research on this topic.

Rosen, M. J. (1978). *Surfactants and Interfacial Phenomena*. Wiley, New York. Describes the characteristics and properties of a very wide range of commercially available surfactants, providing background theory, and describing applications.

## REFERENCES

Anon (2003). *Chemical Week*, 20 August.

Barnes, G. T. (1978). *J. Colloid Interface Sci.*, **65**, 566.

Barnes, G. T., Lawrie, G. A., and Walker, K. (1998). *Langmuir*, **14**, 2148.

Bénard, H. (1901). *Ann. Chim. Phys.*, **23**, 62.

Berg, J. C., Acrivos, A., and Boudart, M. (1966a). *Adv. Chem. Eng.*, **6**, 61.

Berg, J. C., Boudart, M., and Acrivos, A. (1966b). *J. Fluid Mech.*, **24**, 721.

Hansen, R. S. and Ahmad, J. (1971). *Prog. Surface Membrane Sci.*, **4**, 1.

Israelachvili, J. N. (1991). *Intermolecular and Surface Forces*, 2nd edn. Academic Press, London.

Lord Rayleigh (1916). *Phil. Mag.*, **32**, 529.

Lucassen-Reynders, E. H. and Lucassen, J. (1969). *Adv. Colloid Interface Sci.*, **2**, 347.

Lucassen, J. and Barnes, G. T. (1972). *J. Chem. Soc. Faraday Trans. I*, **68**, 2129.

Lucassen, J. and van den Tempel, M. (1972). *Chem. Eng. Sci.*, **27**, 1283.

Lunkenheimer, K. and Miller, R. (1979). *Tenside Detergents*, **16**(6), 312.

Marangoni, C. (1871). *Nuovo Cimento [2]*, **16**, 239.

Maragoni, C. (1878). *Nuovo Cimento [3]*, **2**, 97.

Miles, G. D. and Shedlovsky, L. (1944). *J. Phys. Chem.*, **48**, 57.

Mukerjee, P. and Mysels, K. J. (1971). *Critical Micelle Concentrations of Aqueous Surfactant Systems*. Nat. Stand. Ref. Data Ser., Nat. Bur. Stand. (US).

Posner, A. M., Anderson, J. R., and Alexander, A. E. (1952). *J. Colloid Sci.*, **7**, 623.

Preston, W. C. (1948). *J. Phys. Colloid Chem.*, **52**, 84.

Princen, H. M. and Mason, S. G. (1965). *J. Colloid Interface Sci.*, **20**, 353.

Princen, H. M., Mason, S. G., and Overbeek, J. Th. G. (1967). *J. Colloid Interface Sci.*, **24**, 125.

Rodriguez, S. and Offen, H. (1977). *J. Phys. Chem.*, **81**, 47.

Sally, D. J., Weith, A. J., Argyle, A. A., and Dixon, J. K. (1950). *Proc. Roy. Soc. (London)*, **A203**, 42.

Schofield, R. K. and Rideal, E. K. (1925). *Proc. Roy. Soc. London A*, **109**, 57.

Shinoda, K., in Shinoda, K., Nakagawa, T., Tamamushi, B.-I., and Isemura, T., (1963). *Colloidal Surfactants: Some Physicochemical Properties*. Academic Press, New York, p. 7.

Simister, E. A., Thomas, R. K., Penfold, J., Aveyard, R., Binks, B. P., Cooper, P., Fletcher, P. D. I., Lu, J. R., and Sokolowski, A. (1992). *J. Phys. Chem.*, **96**, 1383.

Sinclair, D. and La Mer, V. K. (1949). *Chem. Rev.*, **44**, 245.

Thompson, J. J. (1855). *Phil. Mag.*, **10**, 330.

Vinson, P. K., Bellare, J. R., Davis, H. T., Miller, W. G., and Scriven, L. E. (1991). *J. Colloid Interface Sci.*, **142**, 75.

Ward, A. F. H. and Tordai, L. (1946). *J. Chem. Phys.*, **14**, 453.

Wark, I. W. (1979). *Chem. Aust.*, **46**, 511.

Weast, R. C. (1977). *Handbook of Chemistry and Physics, 58 edn.*, CRC Press, West Palm Beach, p. F-43.

Weil, I. (1966). *J. Phys. Chem.*, **70**, 133.

## EXERCISES

**4.1.** A surfactant solution of sodium dodecyl sulfonate (concentration 1.70 mmol kg$^{-1}$) is found to have a surface tension of 63 mN m$^{-1}$ at 25 °C. Calculate the adsorption of the surfactant at the air–solution interface and state the two assumptions that are required. The surface tension of pure water at this temperature is 72.0 mN m$^{-1}$.

**4.2.** The cmc of hexadecyl trimethyl ammonium bromide in water occurs at 0.96 mmol kg$^{-1}$ and 38.6 mN m$^{-1}$. A solution of concentration 0.56 mmol kg$^{-1}$ has a surface tension of 47.4 mN m$^{-1}$. Calculate the adsorption at a concentration of 0.63 mmol kg$^{-1}$ and state the assumptions made.

**4.3.** Using a Weil bubble apparatus, 2.0 cm$^3$ of collapsed foam were collected after the passage of 12 dm$^3$ of nitrogen gas. The concentrations of surfactant in the collapsed foam and the parent solution were respectively 0.055 mol dm$^{-3}$ and 0.030 mol dm$^{-3}$. Passage of some of the foam through a capillary of diameter 1.6 mm gave an average length between soap films (defining the extent of each bubble) of 9.95 mm. Assuming that the bubbles in the foam are spherical, calculate the adsorption of surfactant at the gas/solution interface.

**4.4.** Starting from the Szyszkowski equation (4.1) derive the Langmuir equation (4.2).

**4.5.** Calculate the theoretical upper limit for the adsorption of butanol at the air–solution interface using the linear plot in Figure 4.6.

**4.6.** Derive Eqs. (4.15) and (4.16) from (4.13) and (4.14).

**4.7.** A monodisperse aerosol of dibutylphthalate is passed into a flask containing a solution of toluene in dibutylphthalate (mole fraction of toluene $= 0.60$), and allowed to come to equilibrium with the vapour from this solution. The mole fraction of toluene in the aerosol droplets is then found to be 0.57. Assuming that Raoult's law applies to the solutions, calculate the radius of the aerosol droplets at equilibrium. Additional data: $T = 298$ K; $\rho_{toluene} = 867$ kg mol$^{-1}$; $\gamma = 32$ mN m$^{-1}$ (for $x_{toluene} = 0.57$); vapour pressure of dibutylphthalate is negligibly small.

**4.8.** Four towns are situated at the corners of a square with sides of 45 km. They are all to be connected by a road, but as roads are expensive to construct, the minimum length of road is desired. Devise various road plans, calculate the road length for each, and compare with the road length using the plan shown in Figure 4.23.

**4.9.** Show that when three identical soap films meet, the angle between them must be 120°.

# 5 Insoluble monolayers and Langmuir–Blodgett films

## 5.1 Introduction

Near the end of the nineteenth century Agnes Pockels (1891), working in the kitchen of her parents' home, discovered how to manipulate oil films on water and thus developed the first surface film balance (Figure 5.1). Her results led Rayleigh (1899) to suggest that the films were only one molecule thick, thus providing the foundations for the study of insoluble monolayers. Further understanding and improved experimental methods can be attributed to Devaux, Hardy, and Langmuir (1917) and the subject blossomed. A comprehensive review of the work up to 1966 has been published by Gaines (1966).

In Section 4.4 we considered the adsorption at the gas–liquid interface of amphiphiles whose hydrophilic-lipophilic balance made them soluble in water. Adsorbed films of such substances have significant dynamic and equilibrium relationships with one of the bulk phases and much information about the adsorbed films can be obtained from these relationships and from measurements on the bulk phase. Similar amphiphiles with a stronger (usually larger) lipophilic or hydrophobic moiety are less soluble in water and in the extreme are practically insoluble. Such water-insoluble amphiphiles have special properties and, in particular, they are able to form insoluble films at the air–water interface. Generally with these films, little useful information can be obtained from their relationships with a bulk phase so attention is focussed on the films themselves. This task is much easier than for adsorbed films as the molecules are restricted to the surface.

In most cases when an insoluble film forms at the air–water interface it is one molecule thick as this arrangement allows the hydrophilic part of the amphiphile to lie in the water with the hydrophobic part out of the water and

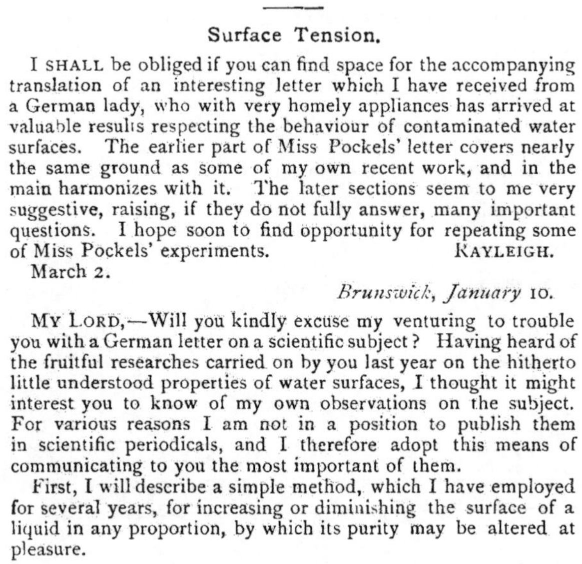

**Figure 5.1.** Extract from *Nature*, 12 March, 1891. We can surmise that the 'various reasons' were mainly, if not entirely, that she was female.

### Surface Tension.

I SHALL be obliged if you can find space for the accompanying translation of an interesting letter which I have received from a German lady, who with very homely appliances has arrived at valuable results respecting the behaviour of contaminated water surfaces. The earlier part of Miss Pockels' letter covers nearly the same ground as some of my own recent work, and in the main harmonizes with it. The later sections seem to me very suggestive, raising, if they do not fully answer, many important questions. I hope soon to find opportunity for repeating some of Miss Pockels' experiments. RAYLEIGH.
March 2.

*Brunswick, January* 10.

MY LORD,—Will you kindly excuse my venturing to trouble you with a German letter on a scientific subject? Having heard of the fruitful researches carried on by you last year on the hitherto little understood properties of water surfaces, I thought it might interest you to know of my own observations on the subject. For various reasons I am not in a position to publish them in scientific periodicals, and I therefore adopt this means of communicating to you the most important of them.

First, I will describe a simple method, which I have employed for several years, for increasing or diminishing the surface of a liquid in any proportion, by which its purity may be altered at pleasure.

in the air. Such a film is called an insoluble monolayer and is formed from substances that are insoluble in both of the adjoining bulk phases (usually water and air). The terms Langmuir film and floating monolayer are also in common use. The lower bulk phase is called the subphase.[4]

If a solid surface is passed through a floating monolayer, the monolayer may deposit on this solid substrate. The process is known as Langmuir–Blodgett or LB deposition after the scientists who discovered and developed the technique. Repeated passages of the substrate through the monolayer may lead to the deposition of additional layers on the surface, thus forming multilayer films. The technique and these LB films will be discussed in detail later in this chapter.

## 5.2 Formation and handling of floating monolayers

### 5.2.1 Molecular requirements

Substances that may form monolayers consist of molecules that are amphiphilic (see Chapter 4). The hydrophobic moiety must be large enough to make the molecule insoluble in water (the usual subphase), while the hydrophilic moiety must have sufficient attraction to water to anchor the molecules to the water surface and prevent them from piling on top of one another.

### 5.2.2 Spreading

In order to form a monolayer on a water surface the amphiphile may be spread in one of two ways:

- Drops or crystals of the amphiphile, when placed on the surface, spread spontaneously to form a monolayer until either the bulk material is exhausted or an equilibrium is reached with the spread monolayer. At equilibrium, the surface pressure (see Eq. 5.2) is the equilibrium spreading pressure, $\Pi^{eq}$.

- The amphiphile is dissolved in a volatile solvent and drops of this solution are placed on the surface. The solvent must have a positive spreading coefficient (see Eq. 2.6 and Table 5.1). The solution spreads, the solvent evaporates, and the monolayer is left. Known amounts of amphiphile may be spread using a measured volume of a solution of known concentration. When the film is to be spread on a small area (as on a surface film balance) very small volumes are needed (50–100 μL) and a microlitre syringe would be used. Commonly used spreading solvents include hexane, benzene, and chloroform, but sometimes it is necessary to increase the solubility of the amphiphile by making the solvent more polar with addition of methanol (<30%).

**Table 5.1.** The initial spreading coefficients of some commonly used spreading solvents on water at 20 °C. Although pure ethanol and methanol are not used as spreading solvents, they are sometimes added to other solvents to improve the solubility of the amphiphile.

| Solvent | Spreading coefficient/ $mN\,m^{-1}$ |
|---|---|
| Hexane | 3.4 |
| Benzene | 8.9 |
| Toluene | 6.8 |
| Chloroform | 13.9 |
| Ethanol | 50.4 |
| Methanol | 50.1 |

4. Although the term 'substrate' is often used, it is better to reserve it for the solid materials on which Langmuir–Blodgett films or self-assembled monolayers are deposited.

As the spreading coefficients of some of these solvents are quite low, it is important that the addition of spreading solution should cease before any appreciable surface pressure develops. If too much solution is deposited the spreading coefficient may become negative and the last few drops will not spread. It will not then be possible to calculate the amount of material spread as a monolayer.

### 5.2.3 The surface film balance

In the laboratory, monolayers are manipulated on a **surface film balance**, sometimes called a **Langmuir trough**. It consists of a shallow trough with hydrophobic edges and hydrophobic barriers (in modern versions, trough and barriers are usually constructed of poly(tetrafluoroethylene), PTFE), and an electronic device is used for measuring surface tension (Figure 5.2).

The trough is slightly overfilled with water and the barrier, lying across the trough, divides the surface and is used to restrict the area of a monolayer spread on the water surface. The monolayer is spread by delivering a measured volume of solution of known concentration. After the volatile solvent has evaporated the monolayer can be compressed or expanded by sliding the barrier along the trough edges. The geometrical area ($A$) of the surface occupied by monolayer and the amount of amphiphile on the surface ($N_M$) give the **area per molecule**:

$$\hat{A}_M = A/N_M. \tag{5.1}$$

The surface tension can be measured either by using a horizontal float barrier to detect the difference in tension between a clean water surface on one side and the monolayer-covered surface on the other side, or by measuring the vertical pull of surface tension on a wetted Wilhelmy plate hanging though the surface (see Section 2.6.1). The result is usually reported as **surface pressure** ($\Pi$), the decrease in surface tension:

$$\Pi = \gamma_w - \gamma_f \tag{5.2}$$

in which $\gamma_w$ is the surface tension of pure water, and $\gamma_f$ is the surface tension of the surface with the film.

It is essential that the contact angle on the Wilhelmy plate be zero. Plates of roughened mica, platinum, or glass have been used in the past, but they must be used with a receding contact angle. Now plates made of paper are favoured as they become saturated with water and both the receding and advancing contact angles are always zero. This enables reliable surface pressures to be

**Figure 5.2.** The basic surface film balance. Surface pressure is measured by the Wilhelmy plate suspended from a force sensor; the monolayer area can be changed by sliding the barrier along the edges of the trough.

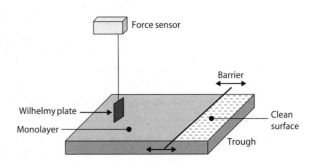

measured when the monolayer is either compressed (receding contact angle) or expanded (advancing contact angle) and gives better control of the surface pressure when that is being held constant for a measurement or a procedure (such as LB deposition).

Most work has been performed with aqueous subphases, but there have been a few experiments on **oil surfaces** (Ellison and Zisman, 1956). The trough was constructed of poly(tetrafluoroethylene) which is both hydrophobic and oleophobic. Surface pressures were measured with an oleophobic horizontal float. Film leakage past this float and also the compressing barrier was controlled by jets of nitrogen gas. Most monolayers in these and later experiments were formed from partly fluorinated compounds or from silicone derivatives as these possessed the necessary insolubility.

### 5.2.4 **The Brooks frame**

A special film balance has also been developed by Brooks and Pethica (1964) to study monolayers spread at the **oil–water** (O–W) or **water–oil** (W–O) interfaces, where the term *oil* refers to a water-insoluble organic liquid. The monolayer is contained between the two top edges of a glass frame and two barriers located at the O–W or W–O interface within a trough. The frame and barriers are rendered hydrophobic. For an O–W interface, the inner sides of the frame and the undersides of the barriers (between the frame edges) are made hydrophilic leaving the top edges of the frame and the other sides of the barriers hydrophobic. For a W–O interface, the top edges of the frame and the edges of the barriers are made hydrophilic. Surface pressure may be measured with a hydrophobic plate which becomes completely coated with the oil giving zero contact angle. Films could be spread at the interface using a long syringe needle and spreading solvents that dissolved in one of the bulk phases.

## 5.3 **Surface pressure–area relationships**

The basic experimental result is an isotherm of surface pressure against area per molecule. From the shape of such isotherms it is possible to recognize four principal monolayer phases: a **gaseous (G) phase**, a **liquid-expanded (Le or L₁) phase**, a **liquid-condensed (Lc or L₂) phase**, and a so-called **solid (S) phase**. These phases are illustrated in the schematic isotherms shown in Figure 5.3. Recent work shows that other phases also exist but their presence is less obvious.

First order transitions between the G and Le or Lc phases are shown as constant surface pressure regions, although in practice the surface pressures are too low to be measured unless specially sensitive equipment is used. The transition between the Le and the condensed phases (Lc or S) is first order although a small rise in surface pressure is usually seen. The transition between the Lc and S phases is second order.

Because of the fairly narrow ranges of temperature and surface pressure available in practice, a particular monolayer may not exhibit all of the features shown in Figure 5.3.

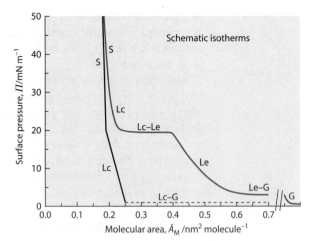

**Figure 5.3.** Schematic $\Pi$-$\mathring{A}$ isotherms showing two common types of behaviour, and the four main phases: S, Lc, Le, and G. Areas are roughly those for single chain amphiphiles. Note that the G phase occurs at much larger areas than shown.

## 5.4 Deposition of Langmuir–Blodgett films

Langmuir–Blodgett films are formed on solid substrates by vertical passage through a floating monolayer. However, not all substrates and not all monolayers are suitable. The choice of substrate is usually determined by the eventual use of the LB film; for example, mica is often used for atomic force microscopy (AFM), silicon for X-ray diffraction, quartz for optical studies and deposited metal films for IR spectroscopy. For satisfactory deposition, the surface pressure should be sufficiently high that the monolayer is in a condensed state. A surface pressure between 20 and 40 mN m$^{-1}$ is normally used.

The efficiency of deposition is measured by the **transfer ratio** which, at constant surface pressure, is the ratio of the area decrease on the subphase surface to the area of substrate apparently covered by the monolayer. With suitable materials, transfer ratios of one are often achieved.

Three types of deposition may be observed: Y, X, or Z. Once the first monolayer has been deposited on the substrate, subsequent movements in and out of the surface often lead to the deposition of an additional layer on each passage. This is known as Y-type deposition: transfer ratios of one on each upstroke and each downstroke. X- and Z-type depositions refer to situations where deposition only occurs on the downstrokes or on the upstrokes,

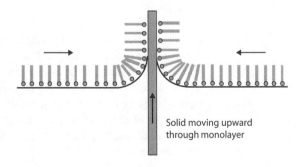

**Figure 5.4.** The Langmuir–Blodgett deposition process.

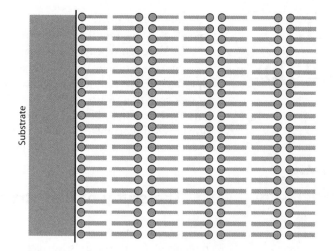

**Figure 5.5.** Probable structure of a multilayer LB film after Y-type deposition on a hydrophilic substrate.

**Figure 5.6.** Possible structures for (a) X-type LB deposition, (b) Z-type LB deposition. However, with some amphiphiles there is evidence that some molecules overturn to give structures similar to that for Y-type deposition (Figure 5.5).

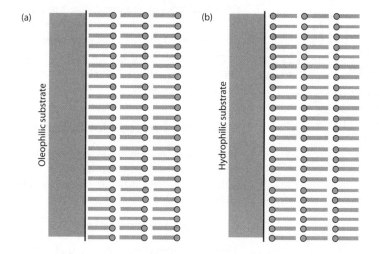

respectively. Note that the definitions of X-, Y-, and Z-type depositions refer to the behaviour during deposition, not on the structure of the film that is formed. In ideal cases, however, these types of deposition are likely to result in the structures shown in Figure 5.5 and Figure 5.6.

Much work has been done using monolayers of carboxylic acids and their salts (Peng *et al.*, 2001). For good-quality films careful preparation of the substrate surface is essential. Selection of the appropriate pH and the type and concentration of salt in the subphase affect the type of film that is formed.

It is important to note that the highly ordered layer structure of most LB films, as illustrated in Figure 5.5, is a major consideration in potential applications.

With very rigid monolayers, a technique developed by Langmuir and Schaefer (1938) can be useful. The monolayer is lifted off the water surface by touching it with a horizontal, hydrophobic substrate, as illustrated in Figure 5.7.

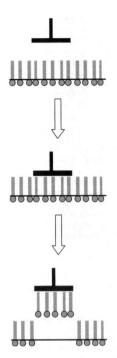

**Figure 5.7.** The Langmuir–Schaefer technique for depositing a monolayer on a solid substrate.

Usually only a single layer is deposited by this **Langmuir–Schaefer** technique, but X-type multilayers may be formed in certain cases (Ulman, 1991, p. 127). Although repeated touchings should give multilayers corresponding to X-type deposition by the LB technique, *some examples* have been reported to produce Y-type monolayers, indicating that the molecules had overturned during deposition (Kato, 1988).

We note here that thin layered structures can also be formed by a process known as **self-assembly**. As this is essentially concerned with adsorption at the solution–solid interface it is discussed in Chapter 9.

## 5.5 The study of film structure

Until recently, the examination of monolayer structure was based on measurements of $\Pi–\hat{A}$ isotherms, surface potentials, and surface viscosities so no direct structural information was available. Now, with great improvements in the sensitivities of a number of techniques and the development of some novel techniques, much detailed structural information can be obtained. Brief descriptions of the various experimental techniques follow. Also included are some techniques that cannot be used directly on floating monolayers, but are applicable to LB films. Sometimes such information can be related back to the parent monolayer.

### 5.5.1 The surface film balance

This is the basic technique for studying floating monolayers. It is also used in the formation and manipulation of monolayers for some of the other techniques described below. It has already been described in Section 5.2.3.

### 5.5.2 Surface potential

There are three electrical potentials that should be identified. The **Volta** or **outer potential** ($\psi$) is the potential just outside a phase boundary: it is the work required to bring unit positive charge from infinity in vacuum to that point. The **Galvani** or **inner potential** ($\phi$) is the work required to bring unit positive charge from infinity into the phase. The potential jump at a phase boundary is denoted by $\chi$. Thus

$$\phi = \psi + \chi. \tag{5.3}$$

The **surface potential** of a monolayer is defined as the difference in the phase boundary potential or Volta potential between a monolayer-covered surface and a clean surface.

$$\Delta E = \psi^{m} - \psi^{o}. \tag{5.4}$$

It usually arises because the hydrophilic part of the amphiphile is either a dipole or an ionized group and thus there is a dipole moment that can substantially modify the electrical properties of the interface.

Surface potentials are usually measured by one of two methods. In both, a potentiometer is normally included in the circuit and adjusted until an electrical balance is obtained. There are considerable problems with shielding in both methods.

- An air electrode carrying a source of $\alpha$ radiation is positioned a few millimetres above the surface, making electrical contact with the surface by ionization of the air gap, and a reference electrode makes electrical contact with the subphase, often via a salt bridge. The potential difference is measured with a high impedance electrometer.

- A **vibrating plate** just above the surface forms an electrical capacitor with the surface and the capacitance changes resulting from the vertical vibrations generate an electric current.

The data enable estimates of the vertical component of the dipole moment to be made and this information is sometimes used to determine the orientation of the hydrophilic moiety. Changes in the surface potential can reveal changes in molecular orientation or conformation resulting from changes in such properties as the packing density.

### 5.5.3 Surface viscosity

The surface shear viscosity of a floating monolayer may be measured by forcing the monolayer to flow through a narrow canal (Jarvis, 1965) or, more usually, by some form of rotating or oscillating probe sited in the surface. A variety of probe geometries have been proposed, but for most it is impossible to make proper allowance for the coupling of the probe and the film to the subphase liquid (Goodrich, 1973). Poskanzer and Goodrich (1975) and Burton and Mannheimer (1967) have designed instruments where the hydrodynamics are sufficiently well defined for the necessary corrections to be made. Goodrich's apparatus consists of a (non-rotating) vertical cylinder with a circular knife edge in the wall. The knife edge is level with the surface and its rotation causes the film to rotate but this rotation is hindered by the drag of the liquid beneath. Talc particles on the surface enable the movement of the film to be measured and the surface shear viscosity to be calculated. The viscous traction instrument of Burton and Mannheimer consists of an outer circular vessel which rotates at constant speed and holds the bulk liquid phase, and two fixed concentric cylinders extending almost to the bottom of the outer vessel. The film to be tested is spread on the annular liquid surface between the two fixed cylinders and its motion is monitored by placing a few hydrophobic particles on the surface (Mannheimer and Schechter, 1970).

The static surface dilational modulus can be calculated from the slope of the $\Pi$–$\hat{A}$ isotherm:

$$K^{\sigma} = -A\left(\frac{\mathrm{d}\Pi}{\mathrm{d}A}\right) = -\frac{\mathrm{d}\Pi}{\mathrm{d}\ln A} \tag{5.5}$$

but this provides little new information. The dynamic modulus is more useful, giving values for the dilational viscosity and elasticity (Lucassen and van den Tempel, 1972). A compression wave is generated by causing a barrier to oscillate back and forth and the amplitude and phase of the resulting surface pressure changes are monitored. The dynamic modulus can be expressed as a complex quantity:

$$
\begin{aligned}
K^\sigma &= |K^\sigma| \exp(i\theta) \\
&= |K^\sigma|(\cos\theta + i\sin\theta) \\
&= K_r^\sigma + iK_i^\sigma
\end{aligned}
\tag{5.6}
$$

where $\theta$ is the phase angle between the surface pressure and the barrier motion after allowing for wave propagation effects. Low-frequency oscillations ($<1\,Hz$) generate compression waves with wavelengths that are usually comparable to the length of the trough and are reflected back from the trough ends. This combination of initial and reflected waves results in uniform oscillations in surface pressure over the whole surface, greatly simplifying the analysis.

Higher frequencies ($>10\,Hz$) generate waves with shorter wavelengths which usually decay before reaching the end of the trough. Thus the amplitude and phase of the observed oscillations in surface tension depend on the position of measurement and appropriate allowances must be made. Frequencies between these two extremes generate complex wave patterns. For accurate work the imperfect transfer of motion from the oscillating barrier to the monolayer must also be considered (Crone et al., 1980).

### 5.5.4 X-ray scattering techniques

Because monolayers and LB films are so thin compared to the substrate on which they are deposited, it is necessary to use X-rays at a very low incident angle and to use very bright sources in order to achieve surface sensitivity and therefore a good signal to noise ratio. Synchrotron X-ray sources are normally used and are essential for diffraction measurements.

In-plane diffraction occurs when there is a two-dimensional lattice structure in the monolayer. In the technique known as grazing incidence X-ray diffraction (GIXD), a diffraction spot is generated when the lattice lines ('planes') fulfil the Bragg criterion for diffraction:

$$
d_{hk} = n\lambda/2\sin\theta
\tag{5.7}
$$

where $d_{hk}$ is the distance between the reflecting lines, $\lambda$ is the wavelength of the X-rays, $n$ is $1, 2, 3, \ldots$, and $\theta$ is the angle in the surface plane between the beam and the reflecting lines. The diffracted beam makes an angle $2\theta$ to the incident beam so, as this angle can readily be measured, $d_{hk}$ can be determined.

With floating monolayers the diffraction pattern is detected by scanning with a vertically oriented position sensitive detector mounted behind a Soller collimator. The low angle of incidence means that there is a long footprint of

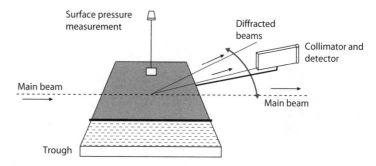

**Figure 5.8.** Apparatus for the measurement of X-ray diffraction from a floating monolayer. Because of the very low incidence angle the footprint of the beam on the monolayer is quite long, so the collimator is used to select diffraction from only a small length of the footprint. The collimator and detector scan in an arc around this point.

the beam on the surface so the collimator is needed to ensure that only radiation coming from the surface at the selected angle enters the detector (Figure 5.8). With LB films the sample can be made sufficiently small to remove any ambiguity about the source of the diffracted radiation so the collimator is unnecessary and area detection methods can be used (Foran *et al.*, 1996). An area detector can be an imaging plate (a plate coated with a material that is changed by X-rays) or a CCD array.

Rotation of the monolayer under the X-ray beam does not change the diffraction pattern so some portion of the monolayer must always be correctly aligned for diffraction. Hence the monolayer must be like a two-dimensional powder, with randomly oriented crystalline domains. If the long chains of the amphiphile are not vertically oriented, the diffraction does not occur in the film plane but is shifted slightly. Measurement of this shift allows calculation of the tilt angle of the chains.

X-ray reflectometry refers to the measurement of the intensity of a beam specularly reflected from the surface at angles above the critical angle for total external reflection. It arises from changes in the electron density in the vertical direction (out of the surface plane). Either the wavelength or the incident angle may be scanned and the changes in reflectivity are then compared with the theoretical pattern generated from a layer model of the interface. Information about the thickness and electron densities of each layer in the appropriate model is obtained. It is thus possible, for example, to extract information about the stoichiometry of a fatty acid monolayer on a solution containing divalent ions.

Wavevector transfer, $Q_z$, is the usual function for reporting the scattering angles in the $z$ direction as it allows for scans by wavelength or angle of observation. It is given by

$$Q_z = (4\pi/\lambda)\sin\varphi \tag{5.8}$$

where $\lambda$ is the wavelength and $\varphi$ is the scattering angle in the vertical ($z$) direction. An example is shown in Figure 5.9.

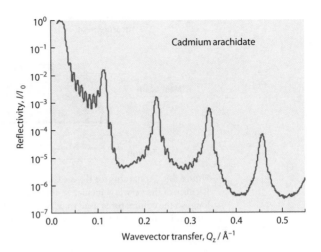

**Figure 5.9.** X-ray reflectivity scan from a multilayer film of cadmium arachidate. The large peaks arise from the repeated bilayers with a spacing of 5.5 nm, while the smaller oscillations between them are a consequence of interference from the top of the multilayer and the multilayer–substrate interface, a distance of approximately 64 nm for this sample. (Data of J. Ruggles.)

### 5.5.5 Neutron scattering

With a neutron beam the scattering centres are the atomic nuclei. The scattering power of an atomic nucleus depends on its scattering length, a property that varies from one element to another and even from one isotope to another. Of particular significance are the very different values of scattering length for hydrogen and deuterium. It is therefore possible to replace hydrogen by deuterium in selected molecular species or in selected parts of the one molecular species and thereby highlight the deuterated part, a technique known as contrast variation. It is also possible, by mixing $D_2O$ and $H_2O$, to match the scattering length density of the subphase to that of air and thus eliminate reflection from the air–subphase boundary.

Because of the weakness of available neutron beams, diffraction techniques are not useful, but neutron reflectometry is a very useful method. As with X-rays, the wavelength or the incident angle is scanned and the reflectivity pattern compared with that of a likely layer model. The thickness and the scattering length density of each layer are obtained.

### 5.5.6 Vibrational and electronic spectroscopy

UV–visible spectroscopy is only feasible for monolayers of molecules with very high extinction coefficients. In the apparatus developed by the Möbius group, absorption of light reflected from the surface is measured: the incident light is led to the surface through a fibre optic bundle and the reflected light is collected by a second fibre optic bundle and led to a detector. Usually both the incident and reflected light beams are normal to the surface with the fibres of the two bundles intermixed at the ends above the surface. The bottom of the trough is blackened to minimize unwanted reflections.

Fourier transform infrared spectroscopy for floating monolayers has been developed particularly by Dluhy and associates (1995) as a reflection–absorption technique, FTIR-RAS. The IR beam (which may be polarized) is incident on the surface at an angle between 30° and 60° to the normal and the beam reflected at the same angle is measured. Good signals from the

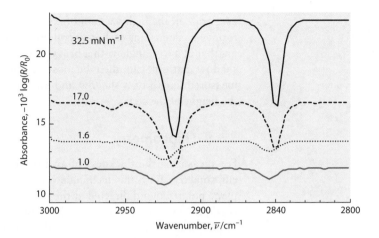

**Figure 5.10.** FTIR-RAS spectra of the C–H stretching vibration region of a floating monolayer of pentadecanoic acid at the surface pressures shown and 292 K, using unpolarized IR radiation at an incidence angle of 30°. (Data of Sinnamon *et al.*, 1999.)

C–H stretching bands are usually observed (see Figure 5.10), but bands arising from the head group are often obscured by absorption bands from the water vapour in the optical path. This latter problem can be overcome by polarization modulation (Blaudez *et al.*, 1996).

As these are reflection–absorption spectra, negative absorbance values are often observed.

With LB films, attenuated total reflection (ATR) and grazing incidence reflection (GIR) are often used. For ATR, the LB film is deposited on both sides of a suitable crystal (germanium or silicon, for example) and the IR beam is directed into the crystal so that it is reflected from both sides a number of times before emerging at the other end and entering the detector. At each reflection the evanescent field of the IR beam enters the deposited film and absorption may occur. A highly reflective metal substrate is required for GIR. Light polarized with its electric vector in the incident plane (*p*-polarized) detects vibrations with a component of the transient dipole moment perpendicular to the metal surface. GIR is thus sensitive to the orientation of molecular groups in the LB film.

### 5.5.7 Electron diffraction

Transmission electron diffraction can be used to study the in-plane structure of LB films. However, as the electron beam interacts strongly with matter, the film must be deposited on a very thin substrate which must also be homogeneous and amorphous. Damage caused by the electron beam is inevitable so exposure times must be short.

### 5.5.8 Electron microscopy

The techniques for imaging monolayers by electron microscopy were developed by Ries (1961). First, the floating monolayer is deposited onto a solid substrate (a coated electron-microscopy grid), usually by the Langmuir–Blodgett technique. As the monolayer material would usually tend to evaporate in the high vacuum of an electron microscope, a carbon replica is formed by depositing carbon vapour in a moderate vacuum and then shadowing that at a low angle with a heavy metal (Pt or Au, heated in vacuum). The shadow pattern can then be observed in the electron microscope. The measured length of a shadow and knowledge of the shadowing angle give the height of the step.

### 5.5.9 Ellipsometry

For measuring the thickness of an LB film, a long-established technique is ellipsometry. With this technique a beam of monochromatic (laser) light is plane polarized and directed at the surface being studied. The reflected light is elliptically polarized because the parallel (s-polarized) and perpendicular (p-polarized) components of the incident light are reflected differently. A compensator changes this elliptically polarized light back into plane polarized light and the angle of polarization is then determined with an analyser and detector. Comparison of the angles of polarization for the incident and reflected light yields a value that depends on both the thickness of the film and its refractive index. Thus a value for the refractive index must be obtained or assumed before the thickness can be calculated.

With floating monolayers, recent versions of the instrument are able to focus on a very small area of surface and so detect certain variations in monolayer structure. For example, the first order nature of the $G \leftrightarrow Le$ and $Le \leftrightarrow Lc$ phase transitions has been confirmed from the coexistence of the two relevant phases (Rasing et al., 1988).

### 5.5.10 Brewster angle and fluorescence microscopy

A p-polarized light beam incident on the surface of water at the Brewster angle (about 53°) is not reflected, but if a monolayer is spread on the surface it changes the refractive index and some reflection occurs. This principle is used in Brewster angle microscopy (BAM) to observe microscopic structures in floating monolayers (Hénon and Meunier, 1991). It is desirable to have contrast between areas where condensed monolayer domains are present and areas where there is relatively little monolayer material. Thus the technique is mostly used in the transition regions between expanded and condensed phases. Interesting domain structures may be observed such as those in Figure 5.11.

Similar information may be obtained by fluorescence microscopy (Lösche et al., 1983). A small quantity (<2%) of a highly fluorescing amphiphilic dye is mixed with the monolayer material and spread. The solubility of the dye should be different in different phases so that when the monolayer is illuminated by a suitable intense light source a difference in fluorescence intensity can be seen through a microscope. In the Le to Lc transition, for example, the Lc phase appears as dark domains in a brightly fluorescing background (Le phase). There is the possibility that the dye could alter the patterns or phase transitions but variations in dye content and comparisons with BAM data show that this is not usually a problem.

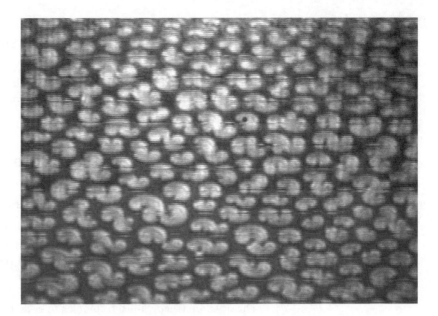

**Figure 5.11.** Brewster angle microscopy image of a monolayer of dipalmitoyl phosphatidyl choline in the transition region between the Le and Lc phases. (From G. A. Lawrie.)

### 5.5.11 Scanning probe microscopy

In scanning probe microscopy the surface is scanned by moving a very fine probe tip over the surface of the sample in a raster pattern. The interaction between probe and sample is measured and the data collated as an image of the surface. In optimum conditions the techniques approach molecular resolution, but are only able to scan a very small sample of the surface ($100 \, nm^2$ to $1 \, mm^2$). Scanning tunnelling microscopy (STM) requires a conducting substrate or film, but atomic force microscopy (AFM) does not have this limitation. These techniques can give useful information about the surface of an LB film, and are described in more detail in Section 7.3.2. They do not require a vacuum and can operate in a range of environments.

### 5.5.12 Surface plasmon resonance

Surface plasmon resonance uses the evanescent wave formed when polarized monochromatic light is totally internally reflected at a metal-coated surface between an optically dense medium and a less dense medium (the sample). At a sharply defined angle of incidence the evanescent wave will interact with free oscillating electrons (plasmons) in the metal leading to the dissipation of energy and a decrease in the intensity of the reflected light. By scanning the incidence angles, the resonance angle is found. This angle depends on the refractive index of the other, less optically dense, medium in the thin region ($<300 \, nm$) penetrated by the evanescent wave. The equipment can thus be used to detect changes in the composition of this thin layer.

### 5.5.13 Miscellaneous techniques for solid surfaces

A number of techniques for studying the surfaces of solids are described in Chapter 7. Of these, some are able to provide useful information about LB films: in particular XPS (X-ray photoelectron spectroscopy) and LEED (low energy electron diffraction).

## 5.6 The structure and properties of monolayers

### 5.6.1 The principal monolayer phases

As indicated earlier, there are four main phases that can be identified from $\Pi$–$\text{Å}$ isotherms, but recent information, mostly derived from GIXD measurements, shows that there is also a rich polymorphism within the condensed phases (S and Lc). Similarities between these polymorphic phases and certain crystalline and smectic phases have been identified (Table 5.2).

The situation is generally too complex to be described here in detail, so only brief outlines are given. Most of the structural information has been obtained for linear amphiphiles (alkanoic acids, alkanols, etc.) as there is, to date, little work reported for more complex molecules. Figure 5.12 is a schematic diagram showing the structures of these main phases. Before giving a more detailed description, it is helpful to understand the **smectic phases** of liquid crystals, as we will see that there are many similarities between smectic phases and monolayer phases.

### S phase

This phase is sometimes referred to as the **solid phase**. It is characterized by a very steep linear isotherm at low areas indicating low compressibility. For simple amphiphiles, such as the straight-chain alkanoic acids and alkanols,

**Table 5.2.** The major monolayer phases and the corresponding structures. Data from Bibo *et al.* (1991).

| Phase | Name | Structure | Smectic category | In-plane area per chain/nm$^2$ |
|-------|------|-----------|------------------|----------------------------------|
| Le, L$_1$ | liquid-expanded | random, vertical or tilted | A or C | $>0.4$ |
| Lc, L$_2$ | liquid-condensed | rectangular (rotator), NN | I | 0.198 |
| Lc′, L$_2$′ | liquid-condensed | rectangular (rotator), NNN | F or H | 0.198 |
| LS | super-liquid | hexagonal (rotator), V | BH | 0.198 |
| S | solid | rectangular (herringbone), V | E | 0.192 |
| CS | closed-packed solid | rectangular (herringbone), V | Xtal | 0.186 |

NN: chains tilt towards nearest neighbour; NNN: chains tilt towards next nearest neighbour; V: chains are normal to the film surface.

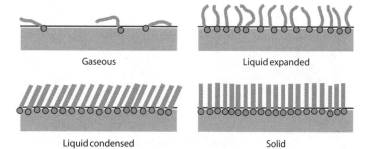

**Figure 5.12.** Schematic diagrams of the main monolayer phases.

Gaseous

Liquid expanded

Liquid condensed

Solid

---

### Smectic phases

Smectic phases, together with nematic phases, characterize the group of materials known as **liquid crystals**. As the name suggests, these are phases intermediate between liquids and crystals and thus are also referred to as mesophases (from Greek: *mesos* = middle). They do not have the highly ordered structure of a crystalline solid, but they do have a greater degree of order than liquids. In particular, they have long-range orientational order. Molecules that form smectic liquid crystals are rod-like and anisotropic.

In smectic liquid crystals the molecules are arranged in layers so it is clear that those spread monolayers where the molecules are oriented in the same direction would fulfil this requirement. The way the molecules are arranged in the film plane determines the class of liquid crystal. The relevant phases are described below.

Smectic A: A fluid phase with the linear molecules normal to the plane of the film (henceforth termed *vertical*) and arranged randomly.

Smectic C: As for smectic A but with the molecules tilted in the same direction.

Smectic BH: Also called the hexatic phase, the molecules are vertical and packed in an hexagonal pattern, but with only short-range positional order.

Smectic I: As for smectic BH, but tilted toward the nearest neighbour.

Smectic F: As for smectic BH, but tilted toward the next-nearest neighbour.

Smectic E: A crystalline phase with vertical molecules arranged in a herringbone pattern (i.e. the planes of the hydrocarbon zigzag are perpendicular to those of the nearest neighbours (see Figure 5.14)).

Smectic K and H: These are tilted crystalline phases, but otherwise like smectic E; K tilted towards the nearest neighbour, H to the next-nearest neighbour.

---

the hydrocarbon chains ('tails') are fully extended (all *trans* bonds) in a planar zigzag of the carbon atoms repeating at 0.254 nm intervals, with the long axis perpendicular to the interface. The hydrophilic 'head' groups lie in the water phase. Evidence for this structure includes the following observations: alteration of the chain length does not alter the area per molecule; the area per molecule at high surface pressures corresponds to the crystallographic cross-sectional area of a hydrocarbon chain ($0.183 \, nm^2 \, molecule^{-1}$); the thickness

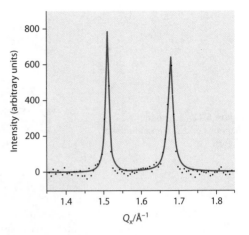

**Figure 5.13.** Synchrotron X-ray diffraction pattern for a monolayer of behenic acid in the S phase at 10 °C and 20 mN m$^{-1}$. The peaks are fitted to a Lorentzian function. (From Barnes *et al.*, 1995.)

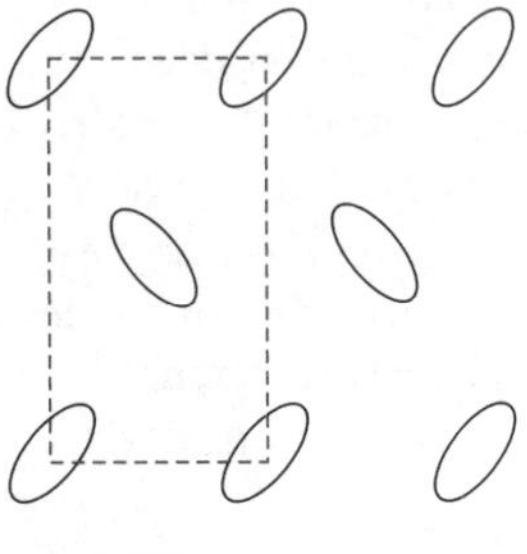

**Figure 5.14.** Suggested in-plane 'herringbone' structure for monolayers of molecules with linear alkyl chains in the S state. Dashed lines indicate the unit cell. Vertical close-packing of chains (ignoring the head groups) in this arrangement gives an area per chain of 0.183 nm$^2$ (Petty, 1996, p. 95).

of the monolayer, determined from the bulk density, agrees with that from electron microscopy and from specular reflection of low angle synchrotron X-rays; and reflectivity measurements with slow neutron beams indicate this orientation. The in-plane structure revealed by grazing incidence synchrotron X-ray diffraction (see, for example, Figure 5.13) is usually a rectangular structure with a herringbone orientation of the alkyl chains (Figure 5.14). Brewster angle microscopy and the fact that the X-ray diffraction pattern is independent of the horizontal direction of the X-ray beam show that the monolayer is an array of randomly oriented crystalline domains, like a two-dimensional powder.

### LS (super liquid) phase

This phase can be considered as an expansion of the S phase sufficient to allow chain rotation about the long axis and a slight change in packing from rectangular to hexagonal (see Table 5.2). It corresponds to the smectic BH (hexatic) structure.

### Lc (liquid-condensed) or L$_2$ phase

The isotherm is nearly linear but less steep than that of the S phase and occurs at slightly higher areas. The structure is controversial: one model has the chains in a somewhat disordered, liquid-like, arrangement; another has the chains fully extended, as in the S phase, but tilted at an angle to the vertical. Recent synchrotron X-ray measurements on some alkanoic acids and alkanols favour the latter, indicating rectangular structures with the chains tilted either to the nearest neighbour (Lc or L$_2$) or the next nearest neighbour (Lc′ or L$_2$′). It appears to be a rotator phase: chains nearly fully extended with no restrictions on the orientation of the zigzag plane relative to the crystallographic axes.

### Le (liquid-expanded) or L$_1$ phase

This phase occurs at larger areas, approximately twice those of the condensed states described above. It is usually formed by molecules that have some impediment to close packing or relatively weak chain–chain attraction (e.g. branched chains, shorter chains, higher temperatures). The chains are disordered, as in a liquid, and transitions from condensed phases to the Le phase are sometimes described as chain melting. The isotherms can be described by an equation of state similar to the van der Waals equation:

$$(\Pi - \Pi_0)(A - A_0) = NkT \tag{5.9}$$

where $\Pi_o$ and $A_o$ are constants. Although some theoretical significance may be attached to these two constants, often in fitting experimental data they are best treated as adjustable parameters.

Alternatively the Amagat equation, as used by Schofield and Rideal (1925), may give a reasonable fit to experimental data:

$$\Pi\left(\hat{A} - \hat{A}_0\right) = qkT \tag{5.10}$$

where $q$ $(<0)$ is a constant and $\hat{A}_0$ gives an area per molecule that corresponds to the effective area occupied by a molecule with its head groups anchored to the surface and the chains, somewhat bent due to gauche conformers, occupying rather larger areas.

### G (gaseous) phase

This phase is rarely observed as it requires equipment of higher sensitivity than that usually available. It occurs at very large areas and at extremely low surface pressures. The molecules are widely separated and move independently like a two-dimensional gas. The equation of state for an ideal gaseous monolayer is reminiscent of the ideal gas equation for three dimensions:

$$\Pi A = NkT \tag{5.11}$$

but a modified form applies in most cases:

$$\Pi(A - A_0) = NkT. \tag{5.12}$$

A calculation based on Eq. (5.11) indicates that at surface pressures about $0.5\,\mathrm{mN\,m^{-1}}$ the area would be about $8\,\mathrm{nm^2\,molecule^{-1}}$, about 40 times the area of a condensed monolayer of a long-chain alcohol or acid $(\approx 0.2\,\mathrm{nm^2\,molecule^{-1}})$.

## 5.6.2 Phase transitions

First order transitions between phases should appear as constant pressure regions in the surface pressure–area isotherm. Transitions between the G phase and the Le, Lc, or S phases do appear to show constant surface pressure, although the accuracy of such measurements is limited by the very low surface pressures involved. In contrast, the Le to Lc and Le to S transitions generally show an appreciable rise in surface pressure during compression. The latter transitions are now known to be first order with the rise in surface pressure attributed to impurities (Hifeda and Rayfield, 1992). The transition from Lc to S is usually a second order transition.

Brewster angle and fluorescence microscopy of the transitions between the G state and other states and between the Le and Lc or S states show domains of the more condensed phase in a continuum of the G or Le phase. In some cases these domains have very complex but uniform shapes. Figure 5.11 is an example showing such domains for a phospholipid monolayer. A two-dimensional foam may sometimes be seen in transitions from the gaseous state. The coexistence of two states through a transition region indicates that the transition is first order, even though the surface pressure does not remain constant (the Le $\leftrightarrow$ S transition, for example).

The surface pressure at which a first order transition occurs depends on the type of molecule, the alkyl chain length in an homologous series, and the temperature. Surface pressure–area isotherms showing a first order phase transition are very similar to pressure–volume isotherms for bulk systems, as Figure 5.15 indicates.

For the mostly second order transition from the Lc to S states the dependence on temperature may be more complex due to subtle changes in the nature of the states involved. For example, the prominent kink in the $\Pi$–$\hat{A}_M$ isotherm

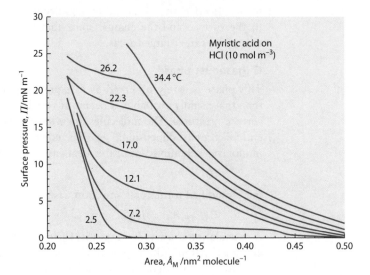

**Figure 5.15.** The effect of temperature (values shown on graph) on the Le to C phase transition for myristic acid. (Replotted from data of Adam and Jessop, 1926.)

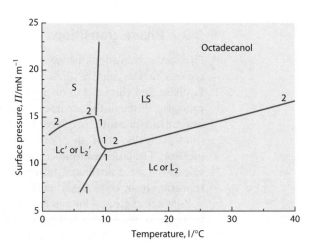

**Figure 5.16.** Phase diagram for octadecanol monolayers. Numbers indicate the order of the phase transition between the regions. (Data of D.S. Hunter, 1990; G.A. Lawrie, 1994.)

of octadecanol rises with increasing temperature except between 8 and 10 °C where there is a sharp fall (Figure 5.16). Similar patterns have been observed with other alcohols and with carboxylic acids.

### 5.6.3 Equilibrium spreading pressure

Economical spreading of a monolayer over a large area, as when long-chain alcohols are used to reduce evaporation from water storages, usually requires that the monolayer be spread without the use of solvents. We noted earlier that when bulk monolayer material (liquid or crystal) is placed on a water surface it will spread spontaneously to form a monolayer. This process continues until either the bulk material is exhausted or equilibrium is reached between the monolayer and the bulk material. The equilibrium position is characterized by an equilibrium spreading pressure, $\Pi^{eq}$, and an equilibrium spreading concentration, $\Gamma^{eq}$. The latter is simply the inverse of the molecular area at the equilibrium spreading pressure.

The equilibrium spreading pressure is dependent on the phase of the bulk solid and the temperature. Each phase may show a different temperature dependence. The existence of crystalline hydrates affects the phase transitions in long-chain alcohols and hence the spreading behaviour (Brooks and Alexander, 1962).

For the temperature dependence of the equilibrium spreading pressure a two-dimensional form of the Clapeyron equation has been proposed by Alexander and Goodrich (1964):

$$\frac{d\Pi^{eq}}{dT} = \frac{d\gamma^{\bullet}}{dT} - \frac{d\gamma}{dT} = -\frac{\Delta H_f - \Delta H_f^m}{TA} \tag{5.13}$$

where $\Delta H_f$ and $\Delta H_f^m$ are respectively the enthalpies of formation of clean surface and monolayer-covered surface from the bulk constituents.

The rate of spreading from bulk material onto the surface is determined by the length of the line of contact between the bulk material and the water surface ($l$) and on the difference between the equilibrium spreading concentration and the actual concentration on the water surface near the bulk material:

$$\frac{dN}{dt} = k^{sp}l(\Gamma^{eq} - \Gamma) \tag{5.14}$$

where $N$ is the number of molecules on the surface. The spreading rate constant, $k^{sp}$, depends on temperature and also on the phase of a crystalline solid, and on which face of a crystal is exposed to the water surface.

Once spread, the monolayer material has to move across the surface and this process can be rapid or extremely slow. Laboratory measurements by O'Brien and Feher (1975) showed rapid movement with hexadecanol and oleic acid, whereas Peng and Barnes (1990) have reported steep gradients in surface pressure with monolayers of several polymers.

### 5.6.4 Monolayer collapse

Continued compression of a monolayer will eventually lead to a situation where material is forced out of the surface and collapse is said to occur.

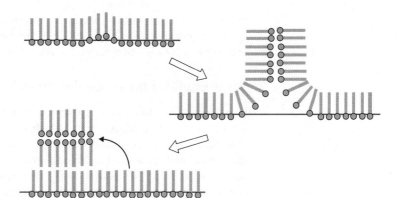

**Figure 5.17.** The mechanism of monolayer collapse suggested by Ries on the basis of electron micrographs showing long narrow structures lying on top of the monolayer with a height equal to two molecular lengths (Ries, 1961).

Fracture and nucleated collapse can be distinguished. Fracture collapse is characterized by a sharp decrease in surface pressure or monolayer area and the surface pressure at which it occurs depends to some extent on the rate of compression. Nucleated collapse, on the other hand, occurs at lower surface pressures and involves a slow but accelerating downward drift in surface pressure or surface area.

When the bulk phase of a monolayer is liquid, it is generally not possible to compress the monolayer above its equilibrium spreading pressure, with droplets of bulk liquid being formed on the surface at this point. However when the bulk phase is a crystalline solid the monolayer may usually be compressed to surface pressures well above the equilibrium spreading pressure. At these high pressures the monolayer is in a metastable state, but there is considerable activation energy for the transition from monolayer to collapsed bulk solid.

On the evidence of electron micrographs of collapsed monolayers of tricontanoic acid, Ries (1961) suggested the mechanism for collapse that is illustrated in Figure 5.17. It is likely that the collapse observed is fracture collapse. Other mechanisms are thought to operate in nucleated collapse (Vollhardt, 1993).

Following rapid compression to a surface pressure below that required for nucleated collapse, there is often a small drop in surface pressure as the molecules readjust to a more stable arrangement.

## 5.7 Interactions in monolayers

### 5.7.1 Two water-insoluble components

The study of monolayers formed by spreading a solution containing two monolayer-forming substances has an interesting history. Some of the earlier experiments were made on mixtures of phospholipids and cholesterol where it was found that the area of the mixed film was less than the sum of the areas for the two separate components. There was a great deal of speculation about the significance of the so-called *condensing effect* of cholesterol on phospholipid monolayers, heightened, undoubtedly, by the presence of both substances in biological membranes. However, such behaviour is not uncommon (another example is shown in Figure 5.18) and it is in fact the more common type of deviation from ideal mixing.

It is considered that if the two components mix ideally, forming an ideal two-dimensional solution, the area should follow the relationship:

$$\hat{A}_{AB} = x_A \hat{A}_A + x_B \hat{A}_B, \tag{5.15}$$

where A and B refer to the two components and AB refers to the mixed film. In this case the two components mix without any change in total area and a plot of area against mole fraction would be linear. Of course, this would also happen if the components were totally immiscible and sometimes it is difficult to distinguish between these two possibilities.

In most mixtures the areas do not follow Eq. (5.15) indicating that there is at least partial miscibility of the components (see Figure 5.18, for example). An excess area of mixing can be calculated for such cases.

By integrating under the $\Pi$–$\hat{A}$ isotherm one obtains the free energy of compression of the monolayer:

$$\Delta G_A = \int_0^\Pi \hat{A}_A d\Pi. \tag{5.16}$$

We note that this equation is based on the alternative definitions of enthalpy and free energy that include the surface term (mentioned after Eq. 3.16). This relation can be used to determine the excess free energy of mixing of the two monolayer components.

If two monolayers are spread on a film balance so that they are separated by a flexible impenetrable thread they will be at the same surface pressure. Removing the thread would, in principle, allow the monolayers to mix and the mixed monolayer could then be brought back to the original surface pressure by adjusting the total area. The free energy change for this adjustment would be the free energy of mixing. In practice, however, this mixing process would be extremely slow so a hypothetical alternative has been suggested by Goodrich (1957). The two components separated by a thread

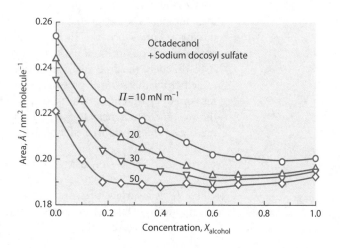

**Figure 5.18.** Molecular areas for mixed monolayers of octadecanol and sodium docosyl sulfate on 0.1 mol dm$^{-3}$ KCl at 298 K. (Data of I. S. Costin.)

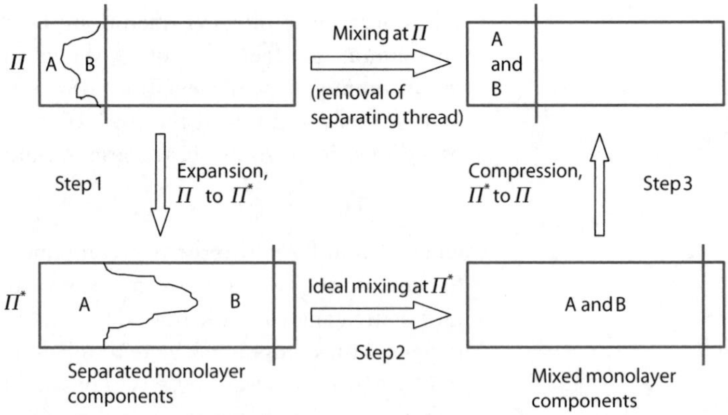

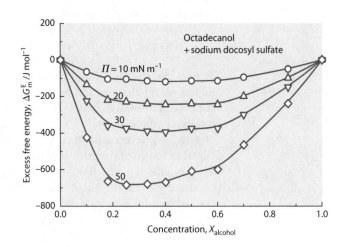

**Figure 5.19.** Plan views of a Langmuir trough showing conceptual operations for the Goodrich method for calculating excess free energies of mixing for monolayers A and B at surface pressure $\Pi$.

**Figure 5.20.** Excess free energies of mixing at the values of $\Pi$ shown, calculated from $\Pi$–$\hat{A}$ isotherms according to Eq. (5.14) for the octadecanol + sodium docosyl sulfate system. (Data of I. S. Costin.)

are expanded from the original surface pressure to a pressure low enough for ideal mixing to occur, probably in the gaseous state (step 1, Figure 5.19). Ideal mixing occurs when the thread is removed (step 2) and then the mixed monolayer is compressed back to the original surface pressure (step 3).

In practice, the free energies of compression using Eq. (5.16) are determined from the surface pressured–area isotherms of the two components, measured separately, and the isotherm of the mixture. From these three values the free energy changes for steps 1 and 3 can be calculated. Step 2 is ideal mixing so the excess free energy change is given by steps 1 and 3. Hence:

$$\Delta G^{E} = \int_{\Pi^*}^{\Pi} \hat{A}_{AB}\, d\Pi - x_A \int_{\Pi^*}^{\Pi} \hat{A}_A\, d\Pi - x_B \int_{\Pi^*}^{\Pi} \hat{A}_B\, d\Pi \qquad (5.17)$$

where $\Pi^*$ is a surface pressure below which the components can be assumed to mix ideally. Data for mixtures of octadecanol and sodium docosyl sulfate are shown in Figure 5.20.

This is a useful procedure for studying interactions in mixed monolayers, but there can be difficulties with the value for $\Pi^*$ as significant non-ideal interactions can occur at surface pressures that are too low to measure accurately.

### 5.7.2 Penetration of monolayers by soluble surfactants

If a water-soluble surfactant is injected into the subphase beneath an insoluble monolayer it may interact with the monolayer in several ways. One possibility is called *penetration* in which the surfactant enters the monolayer between the monolayer molecules and resides in the surface in much the same fashion as the monolayer molecules (illustrated in Figure 5.21).

In a few systems, penetration of the surfactant produces monolayers that are stable up to very high surface pressures, corresponding to very low surface tensions. Under such conditions a surface can expand with the expenditure of very little energy, an observation that led Schulman and his collaborators to the discovery of micro-emulsions (see Section 6.4). Penetration studies also led to a theory for the froth flotation of ores (involving the interaction between frother and collector), and other applications.

Equilibrium studies show that the penetration of the surfactant increases the surface pressure at any selected area per monolayer molecule (such as

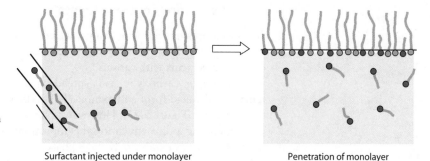

**Figure 5.21.** The process of monolayer penetration: injection of a water-soluble surfactant underneath a water-insoluble monolayer.

Surfactant injected under monolayer          Penetration of monolayer

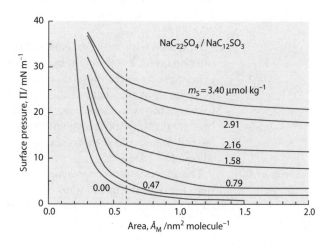

**Figure 5.22.** Equilibrium $\Pi$–$\hat{A}_M$ isotherms for sodium docosyl sulfate monolayers on solutions of sodium dodecyl sulfonate at the concentrations shown. (Data of M. A. McGregor.)

indicated by the dotted line in Figure 5.22). In some systems, compression to low areas appears to completely expel the surfactant and the isotherms merge with the isotherm of the pure monolayer. Mostly, however, the monolayers collapse before this stage is reached (as in Figure 5.22).

The amount of surfactant that has penetrated into a monolayer is not easily determined. A crude measure can be obtained from the increased area at a given surface pressure, and attempts at more accurate measures have mostly been based on the Gibbs adsorption equation (Eq. 3.30). The difficulty with the latter approach is that the surfactant has a significant effect on the monolayer activity, so that the obvious procedure of taking the surface pressures at a selected area per monolayer molecule, such as along the dashed line at $0.6 \, nm^2$ molecule$^{-1}$, and using the simple Gibbs equation does not work. However, these theoretical difficulties appear to have been solved by Hall (1986) (although extra data is required) and it seems that the remaining problems may be attributable to the fact that the Gibbs equation deals with equilibrium adsorption whereas equilibrium is very difficult to achieve with penetration studies, particularly at high surface pressures (Barnes, 2001). Surface analysis by neutron reflectometry or radio-labelling are promising techniques that have not been adequately exploited. However, for penetration studies they do have limitations.

### 5.7.3 Reactions in monolayers

Spread monolayers are able to interact with materials dissolved in the subphase. Possibly the simplest of such reactions is the interaction of monolayers of carboxylic acids with cations.

For example, carboxylic acids spread on a subphase at acidic pH values remain in the acid form and exhibit $\Pi$–$\hat{A}$ isotherms showing both the S and the Lc phases (Figure 5.23). However, if spread on a subphase of higher pH and containing a divalent cation the appropriate salt is formed and the isotherm changes to exhibit only the S phase.

Investigations of the stoichiometry of cadmium stearate monolayers by synchrotron X-ray reflectivity indicate that there is one cadmium ion for each acid ion giving a head group with the formula –COOCdOH (Peng *et al.*, 2000).

Reactions with sections of the hydrophobic part of the monolayer molecule can also occur. For example, the oxidation of double bonds in alkyl chains by permanganate in the subphase first forms the dihydroxy compound followed by splitting of the chain at this point to give two water soluble molecules. Consequently the monolayer area at constant surface pressure increases because the dihyroxy compound attaches itself to the surface at two points instead of one and then decreases as the final products dissolve (Gilby and Alexander, 1956).

The kinetics of such reactions are quite complex as the two reaction steps overlap, the mixed films of acid and dihydroxy compound do not mix ideally (cf. Eq. 5.15) and the movement of soluble products away from the

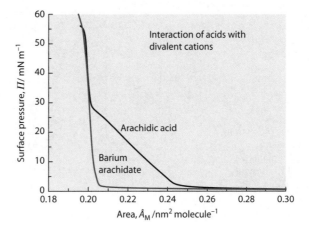

**Figure 5.23.** $\Pi$–$\hat{A}$ isotherms for arachidic acid on an acidic subphase (pH = 6.2) and on an alkaline subphase (pH = 8.2) with 0.5 mol m$^{-3}$ BaCl$_2$. (Data of J. B. Peng.)

surface may be limited by diffusion. Nevertheless, there does appear to be an effect of surface pressure on reaction rate as compression of the monolayer hinders access to the reaction site (e.g. the double bond) on the monolayer molecule.

## 5.8 The structure of LB films

There has been debate in the literature about the relationship between the structure of a monolayer at the air–water interface and the corresponding structure of an LB film formed from its deposition. The relationship is reasonably complex, as it involves variables such as the nature of the subphase, the nature of the amphiphile, the surface pressure at deposition, the number of layers deposited and other factors. In most cases, it is likely that the structure is retained to a large degree upon deposition, but this is not always the case. For example, the first layer may be affected by the underlying substrate structure although subsequent layers are not.

Film structure can be examined by a number of techniques, of which GIXD, AFM, and FTIR-RAS are the most useful.

GIXD gives information about the in-plane structure of the film, the thickness of the film, and the tilting of the alkyl chains. A typical diffraction pattern, obtained using an imaging plate (Foran *et al.*, 1998), is shown in Figure 5.24 and the corresponding scan of in-plane diffraction intensity in Figure 5.25.

Diffraction only occurs when there is a repeating structure in the film, so disordered films or films with a liquid-like molecular arrangement do not give satisfactory signals. It is accepted that the film consists of randomly ordered domains, like a two-dimensional powder. Domain size can be estimated from the width of a diffraction peak, and it is usually found that the domain size is surprisingly small, being only of the order of a few nanometres.

With multilayer films the observation of repeating spots in the z-direction (the vertical rows in Figure 5.24) in GIXD, and the Bragg peaks observable in

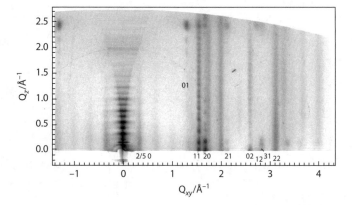

**Figure 5.24.** GIXD diffraction pattern for a cadmium arachidate (CdAr) film of 21 layers (corrected for imaging plate geometry) (Peng *et al.*, 2000).

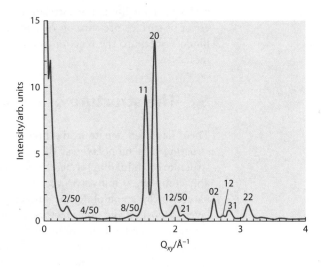

**Figure 5.25.** In-plane diffraction intensity of the CdAr film from data in Figure 5.24. The numbers are the two-dimensional Miller indices (*hk*) for the various diffraction peaks.

reflectometry data (Figure 5.9) demonstrate the layered structure of the films. Generally the repeating structure normal to the film surface has a thickness corresponding to a bilayer, indicating a head-to-head, tail-to-tail arrangement of the molecules as in Figure 5.5.

With some films, GIXD data show diffraction at low scattering angles indicating the existence of superstructures. For example, Peng *et al.* (2003) have found superlattice structures in LB films of cadmium arachidate with changes occurring at about 70, 90, and 103 °C, before a radical transition of the entire film structure at 107 °C from stacked lamellae to hexagonally packed rods.

AFM only gives information about the upper surface of an LB film. It is also very easy to damage the film with the scanning probe so very light probe pressures or a tapping mode are desirable. An AFM image of a monolayer of cadmium arachidate is shown in Figure 5.26.

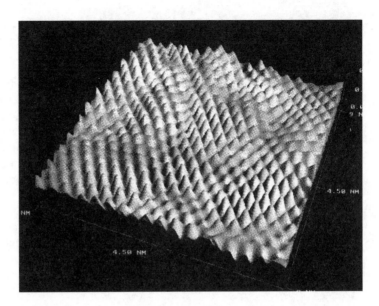

**Figure 5.26.** AFM image of a cadmium arachidate monolayer deposited on mica. (From Peng and Barnes, 1994.)

## 5.9 Applications

### 5.9.1 Spread monolayers

Molecular size, shape, and packing can be determined for suitable compounds.

For example, molar masses of polymers and proteins can be found from $\Pi$–$A$ isotherms of spread monolayers in the gaseous phase by treating the monolayer as an imperfect gas (Eq. 5.10) (see Exercise 5.2).

If the molar mass is known but there are two or more possible conformations of the molecule it is sometimes possible to decide between the possibilities by determining the area per molecule and comparing it with areas determined from models.

#### Membrane modelling

Cell membranes of plants and animals are based on bilayers of amphiphiles which consist of two monolayers back to back. The most common amphiphiles in membranes are phospholipids and there have been many studies of these materials as spread monolayers. A more extended discussion is given in Chapter 10.

#### Lung surfactant

The lungs of animals contain many tiny cavities called *alveoli* that are lined with an aqueous fluid. Breathing involves a substantial increase in the surface area of these alveoli which requires the input of work energy so a reduction of the surface tension of this fluid is essential. This is accomplished in healthy lungs by natural lung surfactant which reduces the surface tension to extremely low values. The absence of this surfactant in some premature infants leads to respiratory distress and, unless treated, to death. Monolayer studies play an important part in the understanding of natural lung surfactant and in the development of exogenous surfactants for therapy. More details are given in Section 10.6.

### Water conservation

Monolayers of cetyl and stearyl alcohols are used to reduce water evaporation rates from dams and reservoirs. In suitable conditions, such as low wind speed and high evaporation rates, the losses by evaporation can be reduced by up to 50%. The alcohols spread readily from small flakes of bulk solid forming condensed monolayers with usefully high equilibrium spreading pressures. However, used by themselves, the alcohol monolayers are easily blown aside by wind, but proposals to incorporate polymeric amphiphiles into the monolayer may produce more stable films.

The ability of a monolayer to retard evaporation is best described by its evaporation resistance, $r$. This may be calculated from measurements of the evaporation rates of water with ($v_f$) and without ($v_w$) monolayer:

$$r = A\Delta c^{eq}\left(\frac{1}{v_f} - \frac{1}{v_w}\right) \tag{5.18}$$

where $\Delta c^{eq}$ is the difference in the equilibrium vapour concentrations of water at the water surface and in the measuring apparatus, and $A$ is the area of surface through which the measured evaporation occurs. In the laboratory, the values of $v_f$ and $v_w$ are readily measured by the apparatus of Langmuir and Schaefer (1943), and a calibration procedure yields the value of $A\Delta c^{eq}$. In this apparatus a powdered solid desiccant is held by a vapour-permeable membrane a few millimetres above the water surface and the change in weight with exposure time is measured.

Recent studies (McNamee *et al.* 1998) indicate that the domains of octadecanol (and probably other alcohols) at high surface pressures are in the highly-ordered S state and do not allow the passage of water. Evaporation occurs through the disordered regions at the domain boundaries.

For a homologous series of alcohols (or acids), the evaporation resistance rises exponentially with chain length, but the rate of spreading from solid particles placed on the water surface decreases. Practical evaporation control therefore is a compromise between evaporation resistance and spreading rate.

This topic has been reviewed by Barnes (1986).

### 5.9.2 LB films

The potential applications of LB films were outlined by Swalen *et al.* in 1987, by Roberts in 1990 (chapter 7) and by Ulman in 1991 (part 5). In the following we briefly outline some of the applications as viewed by Roberts and describe some of the more recent developments.

### Molecular electronics

The continuing miniaturization of electronics with the advantages that it provides for convenience and processing speed has generated considerable interest in the possibilities of molecular electronics. The goal is to produce electronic components such as switches, diodes and transistors using the smallest possible elements: single or very few molecules. LB film deposition is unique in the control offered for producing highly organized molecular arrays of controlled, nanometer-scale thickness, and therefore plays a part in that quest. Interest centres on organic materials with the wide variety of structures that are possible.

This interest has led to the development of Langmuir troughs with various mechanisms for the LB deposition of, for example, alternate layers of two different amphiphiles. Two parallel troughs are used and the substrate for the deposition can be passed either above or below the liquid surface from one trough to the other.

While it is relatively easy to deposit LB films on a metal substrate, the deposition of a second metal layer on top of the LB film usually damages the film. Several strategies for overcoming this problem have been developed and there is now unambiguous evidence for rectification by layers of hexa-decylquinolinium tricyanoquinodimethanide between gold electrodes (Xu *et al.*, 2001). Self-assembled monolayers (see Section 9.11) are another promising way to fabricate devices for molecular electronics (see box).

### Second-harmonic generation

There is considerable interest in extending the frequency range of laser devices. Organic materials can act as highly efficient doublers of the frequency of light (known as second-harmonic generation) in the visible and near infrared regions of the spectrum. In order to achieve second-harmonic generation non-centrosymmetry is necessary, and the LB method is useful as it offers control at the molecular level. Although molecules that are themselves non-centrosymmetric are preferable in some ways, it is not absolutely necessary as the requirement can be satisfied if the *assembly* lacks inversion symmetry. If LB deposition is used, this implies that either X-type or Z-type structures must be achieved, because the inherent centrosymmetry in Y-type structures (Figure 5.5) causes the second harmonic generation (SHG) to be suppressed.

Large SHG is usually associated with molecules that possess the potential for intramolecular charge transfer (Ashwell, 1999) with a long π-bridge between donor and acceptor. When deposited as LB films, such molecules tend to form a centrosymmetric head-to-tail Y-type arrangement, but this can be overcome by interleaving the active layers with inactive spacers. The same end can also be achieved by the use of chromophores with two hydrophobic end groups. In an optimal assembly, the magnitude of SHG shows a quadratic dependence on the number of layers deposited, and this has been achieved to well over 100 active layers.

### Chemical and biological sensors

There are three stages in the operation of a sensor:

- recognition, where there is a specific interaction between the sensor and the target substance;
- a consequent change in some physical property of the sensor;
- conversion of this change into a measurable signal.

In the present context, the sensor would be a LB film deposited on a substrate that forms part of the detection system: such as the surface of a quartz crystal

### Monolayers as diodes: molecular rectification

In the quest for practical molecular electronic devices, several significant barriers must be overcome. One step along the way is to demonstrate the rectification of current by single monolayers of organic molecules. The molecular rectifier is the organic counterpart of the pn junction, and the desired performance can be achieved by a structure formed of self-assembled monolayers with a donor–(electron-bridge)–acceptor sequence which, when sandwiched between electrodes and a voltage applied, permits electrons to preferentially tunnel from the electrode to acceptor at one end of the device and from the donor to electrode at the opposite end. If the current direction is reversed, the flow of electrons is much less, leading to the necessary asymmetrical current–voltage characteristics. Although still a long way from a single molecule device, it represents important progress.

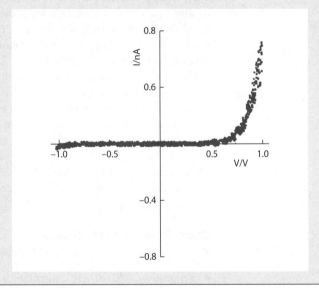

The system used by Ashwell and coworkers to achieve molecular rectification with a self-assembled monolayer (above) and the asymmetric current–voltage (*I–V*) characteristics it displays (below).

microbalance (QCM), an ATR crystal, or a metal film for grazing incidence reflection (GIR). Interaction between the LB film and the target substance would then change the signal from the detection system: the resonance frequency of a QCM, the resonance angle of a surface plasmon resonance apparatus, absorbance of the FTIR signal passing through an ATR crystal or reflected from a GIR surface.

For many applications an LB film may not be sufficiently stable to form a useful sensor as the sample solution is usually required to flow over the test surface. In such cases the stronger bonding of a self-assembled film to its substrate may be needed. Such films are discussed in Chapters 9 and 10.

## SUMMARY

Insoluble monolayers are formed from amphiphiles that are virtually insoluble in water. They can be spread on a water surface from drops of a solution of the amphiphile in a volatile solvent or by spontaneous spreading from crystals or drops of the bulk material. In the laboratory, a surface film balance is used to manipulate monolayers and is a component of various techniques for measuring monolayer properties.

Monolayers exist in a number of phases, most of which can be compared to smectic phases. The major phases are the S, Ls, Lc (liquid-condensed), Le (liquid-expanded), and G (gaseous). The hydrophilic part of each molecule anchors the molecule to the water surface and the hydrophobic part is in the air and may be disordered (Le and G states), or organized into hexagonal or rectangular packing patterns.

Passing a solid substrate through a spread monolayer may result in the ordered deposition of a monolayer onto the substrate, and repeated passages may deposit further layers giving a highly organized layer structure. This is known as Langmuir–Blodgett (LB) deposition.

The structures and properties of spread monolayers and of LB films are described and discussed and the chapter concludes with an outline of actual and potential applications.

## FURTHER READING

Gaines, G. L. (1966). *Insoluble Monolayers at Liquid–Gas Interfaces.* Interscience Publishers, New York. A thorough and comprehensive review of the literature on floating monolayers with a short section on LB films.

Jones, M. N. and Chapman, D. (1995). *Micelles, Monolayers, and Biomembranes.* Wiley-Liss, New York. Chapter 2 gives an overview of monolayer studies with an emphasis on phospholipids.

La Mer, V. K. (ed.) (1962). *Retardation of Evaporation by Monolayers: Transport Processes.* Academic Press, New York. A set of research papers presented at a conference in 1960.

Petty, M. C. (1996). *Langmuir–Blodgett Films: An Introduction.* Cambridge University Press, Cambridge.

Roberts, G. (ed.) (1990). *Langmuir–Blodgett Films.* Plenum Press, New York. A set of authoritative articles covering Langmuir monolayers, LB films, history, and applications.

Ulman, A. (1991). *An Introduction to Ultrathin Organic Films from Langmuir–Blodgett to Self-Assembly.* Academic Press, San Diego. A exhaustive treatment of the subject with many references to original work.

## REFERENCES

Adam, N. K. and Jessup G. (1926). *Proc. Roy. Soc. (London)*, **A112**, 362.

Alexander, A. E. and Goodrich, F. C. (1964). *J. Colloid Sci.*, **19**, 468.

Ashwell, G. J. (1999). *J. Mat. Chem*, **9**, 1999.

Barnes, G. T. (1986). *Adv. Colloid Interface Sci.*, **25**, 89.

Barnes, G. T. (2001). *Colloids Surfaces A*, **190**, 145.

Barnes, G. T., Gentle, I. R., Kennard, C. H. L., Peng, J. B., and Jamie I. McL. (1995). *Langmuir*, **11**, 281.

Bibo, A. M., Knobler, C. M., and Peterson, I. R. (1991). *Makromol. Chem., Makromol. Symp.*, **46**, 55.

Blaudez, D., Turlet, J. M., Dufourcq, C. Q., Bard, D., Buffeteau, T., and Desbar B. (1996). *J. Chem. Soc., Faraday Trans.*, **92**, 525.

Brooks, J. H. and Alexander, A. E. (1962). *J. Phys. Chem.*, **66**, 1851.

Brooks, J. H. and Pethica, B. A. (1964). *Trans. Faraday Soc.*, **60**, 208.

Burton, R. A. and Mannheimer, R. J. (1967). In *Ordered Fluids and Liquid Crystals, Adv. Chem. Series*, Amer. Chem. Soc., Washington, **63**, 315.

Costin, I. S. and Barnes, G. T. (1975). *J. Colloid Interface Sci.*, **51**, 106.

Crone, A. H. M., Snik, A. F. M., Poulis, J. A., Kruger, A. J., and van den Tempel, M. (1980). *J. Colloid Interface Sci.*, **74**, 1.

Dluhy, R. A., Stephens, S. M., Widayati, S., and Williams A. D., (1995). *Spectrochim. Acta*, **A51**, 1413.

Ellison, A. H. and Zisman, W. A. (1956). *J. Phys. Chem.*, **60**, 416.

Foran, G. J., Gentle, I. R., Garrett, R. F., Creagh, D. C., Peng, J. B., and Barnes, G. T. (1998). *J. Synchrotron Rad.*, **5**, 107.

Foran, G. J., Peng, J. B., Steitz, R., Barnes, G. T., and Gentle, I. R. (1996). *Langmuir*, **12**, 774.

Gilby, A. R. and Alexander, A. E. (1956). *Aust. J. Chem.*, **9**, 347.

Goodrich, F. C. (1957). *Proc. 2nd Internat. Congr. On Surface Activity*. Vol. I, p. 85. Butterworth, London.

Goodrich, F. C. (1973). *Progr. Surface and Membrane Sci.*, **7**, 151.

Hall, D. G. (1986). *Langmuir*, **2**, 809.

Hénon, S. and Meunier, J. (1991). *Rev. Sci. Instrum.*, **62**, 936.

Hifeda, Y. F. and Rayfield, G. W. (1992). *Langmuir*, **8**, 197.

Jarvis, N. L. (1965). U.S. Naval Research Laboratory Report 6250.

Kato, T. (1988). *Jpn. J. Appl. Phys. Part 2*, **27**, L1358, L2128.

Langmuir, I. and Schaefer, V. J. (1938). *J. Amer. Chem. Soc.*, **57**, 1007.

Langmuir, I. and Schaefer, V. J. (1943). *J. Franklin Inst.*, **235**, 119.

Lord Rayleigh (1899). *Phil. Mag.*, **48**, 337.

Lösche, M., Sackmann, E., and Möhwald, H. (1983). *Ber. Bunsenges. Phys. Chem.*, **87**, 848.

Lucassen, J. and van den Tempel, M. (1972). *Adv. Colloid Interface Sci.*, **41**, 491.

Mannheimer, R. J. and Schechter, R. S. (1970). *J. Colloid Interface Sci.*, **32**, 195.

McNamee, C. E., Barnes, G. T., Gentle, I. R., Peng, J. B., Steitz, R., and Probert, R. (1998). *J. Colloid Interface Sci.*, **207**, 258.

O'Brien, R. N. and Feher, A. I. (1975). *J. Colloid Interface Sci.*, **51**, 366.

Peng, J. B. and Barnes, G. T. (1990). *Langmuir*, **6**, 578.

Peng, J. B. and Barnes, G. T. (1991). *Langmuir*, **7**, 1749.

Peng, J. B. and Barnes, G. T. (1994). *Thin Solid Films*, **252**, 44.

Peng, J. B., Barnes, G. T., and Gentle, I. R. (2001). *Adv. Colloid Interface Sci.*, **91**, 163.

Peng, J. B., Barnes, G. T., Gentle, I. R., and Foran, G. J. (2000). *J. Phys. Chem.*, **104**, 5553.

Peng, J. B., Foran, G. J., Barnes, G. T., and Gentle, I. R. (2003). *Langmuir*, **19**, 4701.

Peng, J. B., Lawrie, G. A., Barnes, G. T., Gentle, I. R., Foran, G. J., Crossley, M. J., and Huang, Z. (2000). *Langmuir*, **16**, 7051.

Pockels, A. (1891). *Nature*, **43**, 437.

Poskanzer, A. and Goodrich, F. C. (1975). *J. Colloid Interface Sci.*, **52**, 213.

Rasing, T., Hsiung, H., Shen, Y. R., and Kim, M. W. (1988). *Phys. Rev. A* **37**, 2732.

Ries, H. E. (1961). *Scientific American*, March.

Schofield, R. K. and Rideal, E. K. (1925). *Proc. Roy. Soc.*, **A109**, 57.

Sinnamon, B. F., Dluhy, R. A., and Barnes, G. T. (1999). *Colloids and Surfaces A*, **146**, 49.

Swalen, J. D., Allara, D. L., Andrade, J. D., Chandross, E. A., Garoff, S., Israelachvili, J., McCarthy, T. J., Murray, R., Pease, R. F., Rabolt, J. B., Wynne, K. J., and Yu, H. (1987). *Langmuir*, **3**, 932.

Vollhardt, D. (1993). *Adv. Colloid Interface Sci.*, **47**, 1.

Xu, T., Peterson, I. R., Lakshmikantham, M. V., and Metzger, R. M. (2001). *Angew. Chem. Int. Ed.*, **40**, 1749.

## EXERCISES

**5.1.** In 1774, Benjamin Franklin described some experiments he had performed on a pond in Clapham Common, London. He had observed that a teaspoon of oil (5 cm$^3$), placed on the water surface, would calm the surface for up to half an acre (2 × 10$^3$ m$^2$). Calculate the thickness of the film and compare it with the lengths of various long-chain molecules. Suggest reasons why the calming effect did not extend to larger areas.

**5.2.** Calculate the average molar mass and molecular area of egg albumin from the data in the table. The data refer to a monolayer of egg albumin spread on water in a film balance at 25 °C.

**Table** Surface pressure–area data for egg albumin.

| Surface pressure, $\Pi/\text{mN m}^{-1}$ | Area, $A/\text{m}^2\,\text{mg}^{-1}$ |
|---|---|
| 0.07 | 2.00 |
| 0.11 | 1.64 |
| 0.18 | 1.50 |
| 0.20 | 1.45 |
| 0.26 | 1.38 |
| 0.33 | 1.36 |
| 0.38 | 1.32 |

**5.3.** A cholesterol monolayer has an area of 0.405 and 0.40 nm$^2$ molecule$^{-1}$ at 5 and 20 mN m$^{-1}$ respectively and a monolayer of dipalmitoyl phosphatidylcholine (DPPC) has areas of 0.68 and 0.43 nm$^2$ molecule$^{-1}$ at these surface pressures. The corresponding areas for a mixed monolayer consisting of 1 : 3.4 cholesterol : DPPC are 0.38 and 0.40 nm$^2$ molecule$^{-1}$. Calculate the excess areas of mixing at these two surface pressures.

**5.4.** The evaporation resistance of a monolayer is measured with the Langmuir–Schaefer apparatus. Evaporation was measured from an area of 63.6 cm$^2$ and the observed rates were 0.66 and 0.34 mg s$^{-1}$ for the monolayer-free and monolayer-covered water surfaces respectively. Prior calibration had given a value of 16.6 mg dm$^{-3}$ for the difference in the equilibrium concentrations of water vapour over clean water and desiccant. Calculate the evaporation resistance.

# 6 The liquid–liquid interface; membranes

## 6.1 Introduction

Our acquaintance with the liquid–liquid interface begins at a very early age when we have our first taste of mother's milk: an oil-in-water emulsion. As we grow and experience other foods and drinks, the range and variety of liquid–liquid interfaces that we encounter, many of them associated with emulsions, increases enormously. As a consequence the study of food science is closely related to the study of emulsions.

Thus interest in the liquid–liquid interface is largely concerned with the topic of emulsions. These consist of liquid drops (the disperse or internal phase) dispersed in another liquid (the continuous or external phase). The dispersion is usually stabilized by the adsorption of a third component at the liquid–liquid interface.

The movement of a solute from one liquid phase to another is also a topic of considerable practical importance with processes such as liquid–liquid extraction in chemical industry and in many biological situations.

Sometimes the two liquids are miscible, often with the same solvent, and a barrier is required to keep them separate. The barrier may be a membrane or a bridge such as the salt bridges used in electrochemistry. Such barriers keep the two liquids apart while allowing selected components to move through. Bridges are outside the scope of this book. The role of membranes, particularly biological membranes, can be significant in solute transport and is an essential feature of many biological processes. Information relevant to membrane behaviour is often obtained from studies of insoluble spread monolayers, particularly at the organic-liquid–water interface.

## 6.2 Emulsions

Generally in emulsions we are concerned with water or an aqueous solution as one phase and a water-insoluble organic liquid, which can be loosely described as an *oil*, as the other phase. Such emulsions fall naturally into two types: oil-in-water (O/W) and water-in-oil (W/O), where the second liquid forms the continuous phase. In recent times it has become necessary to distinguish between what we might call *normal* or *macroemulsions* and *microemulsions*. While both are emulsions in that one liquid is dispersed in another, there are important differences in their properties that make it desirable to treat them separately. The differences are mainly thermodynamic – despite the terminology which suggests that size is the major determining factor. Thus this section will be concerned with the normal or macroemulsions (called for simplicity, *emulsions*) leaving microemulsions to be treated in Section 6.4.

The term *emulsion* is sometimes used to describe systems where one phase is solid, but was liquid when the dispersion was formed. Cooling or polymerization following the preparation can result in such a system.

Emulsions formed with pure liquids are unstable and quickly separate into different layers. The reasons for this instability are readily apparent from an examination of the interfacial tensions of various oil–water interfaces, as shown in Table 6.1.

The table shows that the interfacial tension is low only when the organic liquid has a polar group that is capable of interacting with water. For the other liquids the relatively high interfacial tension provides a significant driving force for reducing the area of the interface.

Relative stability of an emulsion is only achieved when a third component is present: an emulsifying agent or emulsifier. In broad terms, the emulsifier lowers the interfacial tension and reduces the thermodynamic driving force towards coalescence. The role of the emulsifier in stabilizing an emulsion and in preferentially forming an O/W or a W/O emulsion will be discussed later in this chapter (Section 6.3).

**Table 6.1.** Interfacial tensions of various liquids against water at 20 °C. (Selected from an extensive compilation by Girifalco and Good, 1957.)

| Liquid | Interfacial tension, $\gamma$/mN m$^{-1}$ |
|---|---|
| Hexane | 51.1 |
| Decane | 51.2 |
| Benzene | 35.0 |
| Toluene | 36.1 |
| Carbon tetrachloride | 45.0 |
| Chloroform | 31.6 |
| Oleic acid | 15.7 |
| Octanol | 8.5 |

### 6.2.1 **The formation of emulsions**

There are two principal methods for forming emulsions: dispersion and condensation. Other methods will also be briefly mentioned.

#### Mechanical dispersion

This procedure involves the shearing of a mixture of the two liquids in a 'colloid mill', of which there are a number of different designs. The process is aided by the addition of a surfactant to the mixture and this also helps to stabilize the final emulsion. Spontaneous or easy dispersion occurs when the interfacial tension has been reduced by carefully selected surfactants to a very low value or even, controversially, to a negative value. In such circumstances very little (or zero) mechanical effort is required to generate an emulsion.

Mechanical dispersion is usually performed in two stages: simple agitation to form a coarse dispersion followed by an intense shearing action to yield the fine dispersion required. During each stage the area of the interface is being increased and this process is significantly aided by lowering of the interfacial tension. This is one of the functions of the emulsifier and is clearly related to the other highly important function of stabilizing the emulsion once it has been formed.

In studies of monolayer penetration by water-soluble surfactants (see Section 5.7.2), Schulman and coworkers found that in certain systems the surface tension dropped to values near zero or even to slightly negative values. While there is considerable argument about these results, there is no doubt that certain combinations of an insoluble monolayer material and a water-soluble surfactant can produce very low surface tensions. When such combinations are used as emulsifiers (for example, by dissolving the surfactant in water and the monolayer material in the oil) emulsions are formed easily and may even appear to form spontaneously giving very stable emulsions (Schulman, 1955, 1961). Usually the emulsion droplets are extremely small (10–50 nm diameter) and are called microemulsions. These are discussed further in Section 6.4.

#### Condensation methods

Condensation occurs when separate molecules of the disperse phase come together to form the emulsion droplets. A common procedure for an O/W emulsion is to dissolve the oil phase in ethanol and add this solution to water. The distribution of the water-soluble alcohol throughout the aqueous phase leaves the oil molecules 'stranded', hydrophobic molecules in a hydrophilic environment, so that when they encounter other oil molecules they tend to aggregate with them (see Section 2.9.3). Again the presence of an emulsifier aids the process and stabilizes the final emulsion.

#### Phase inversion methods

This approach utilizes the tendency of a thin sheet of liquid to break up into droplets. In an emulsion with a high proportion of disperse phase there will be a tendency for the continuous phase to break up leading to phase inversion. Then, provided the emulsifier is capable of stabilizing the new emulsion, a fine dispersion will be formed. Thus, for example, if water is added slowly to an oil containing the emulsifier and subjected to a vigorous shearing action a W/O emulsion will form initially and become quite viscous at about 60% water. After removal from the mill and addition of extra water the emulsion suddenly inverts to a O/W emulsion and the viscosity drops.

Interfacial tension decreases with temperature so emulsification is easier at higher temperatures. Thus if an O/W emulsion is formed a few degrees below the phase inversion temperature (see Section 6.3.5) a fine emulsion is formed. It must then be cooled quickly.

### Electrical methods

If the disperse phase is injected into the dispersion medium through a fine jet which is highly charged, the thread of liquid rapidly breaks up into droplets which, being charged, repel one another. An emulsifier in the medium can then stabilize the resulting dispersion.

## 6.2.2 Emulsion type

To distinguish between O/W and W/O emulsions it is usually sufficient to measure the conductivity of the liquid, as an aqueous continuous phase will normally give higher conductivity than an oil continuous phase. Other properties, such as the miscibility of the continuous phase with water or with oil, or the distribution of a dye soluble only in one phase, may also be used.

In order to deliberately generate a particular emulsion type it is necessary to consider several factors: the volume ratio of oil to water, the choice of emulsifier, and the method of preparation can all affect the outcome. These factors are considered further in the following sections.

## 6.2.3 Size distribution

It is no longer desirable to prescribe a lower size limit for emulsion droplets as newer particle sizing techniques can and do now detect droplets well below the limits of optical resolution.

Unless special care is taken during preparation, the particle sizes in an emulsion are distributed over a significant range. Such a range may be characterized by a *geometric mean* and a *standard deviation* related usually to the diameters of the particles. Other statistical measures such as the *median* and the *mode* may sometimes be useful. Special techniques can be used to form emulsions where all of the particles are the same size or, more precisely, have a very narrow size distribution: these are termed **monodisperse** emulsions.

The techniques for forming monodisperse aerosols have been described in Section 4.11. If the droplets of such a monodisperse aerosol are charged by passage through a corona discharge and then bubbled into the electrically grounded bulk continuous phase, an emulsion is formed that displays the higher order Tyndall spectra characteristic of a monodisperse colloid (Wachtel and LaMer, 1959, 1962).

For ordinary emulsions the size distribution is rarely a normal or Gaussian distribution. Mostly it is a *log-normal* distribution where the *distribution of frequencies* of the various sizes, plotted against the *logarithm of the size* (usually the diameter), is Gaussian.

Because of their small sizes and curved surfaces, emulsion droplets have a slightly higher solubility in the dispersion medium than the same material with a flat interface. This is an extension of the Kelvin equation (2.19) that has not been proven experimentally, but accords with relevant experimental observations. Consequently, if the emulsion is polydisperse the smaller droplets will have higher solubilities than the larger droplets and so will tend to dissolve while the larger droplets tend to grow, a form of Ostwald ripening.

### Measurement of the size distribution

The distribution of droplet size in an O/W emulsion can sometimes be determined by a Coulter counter. In this apparatus the emulsion flows through a small hole in a barrier separating two electrodes and the pulse in electrical conductivity as a particle passes through the hole gives a measure of the particle size. However, the technique usually requires the addition of an electrolyte to provide a conducting medium and this may alter the properties of the emulsion.

The pattern of light scattering varies with the size range of the particles: Rayleigh scattering dominates with particles that are small relative to the wavelength of light; Mie scattering occurs when the particles are of similar size to the wavelength of light. For Rayleigh scattering the intensity of light scattered at right angles to the incident light is measured. This is proportional to the number concentration of scatterers (in this size range each droplet is a single scatterer) so if the concentration of disperse material is known the average droplet size can be calculated. In a variant of this technique the intensity of transmitted light, the *turbidity*, is measured. With larger particles there are multiple scatterers within each particle and optical interference yields a complex pattern of scattering intensity with angle of observation, known as *Mie scattering*. Comparison of the observed pattern with that generated by computer using the Mie theory gives the particle size distribution with good accuracy if the particles are spherical. For monodisperse emulsions the scattering of white light shows the colours of the higher order Tyndall spectrum (see Section 9.8.2) with alternating red and green bands dominating. The angles at which these bands are observed yield the particle size.

Other techniques include hydrodynamic methods, particularly field flow fractionation; and electroacoustics.

The topic has been thoroughly reviewed by Hunter (1993).

### 6.2.4 Emulsion stability

The study of emulsion stability has a long history which has been detailed by Becher (2001). The topic is closely related to the selection of the emulsifying agent which will be discussed in more detail in Section 6.3. We will here outline the terminology and the basic processes involved in emulsion stability and breaking.

## Creaming

Flocculation is said to occur when the emulsion droplets cluster together without merging. Often this arises from the usual density difference between oil and water causing the oil droplets in an O/W emulsion to tend to rise to the surface: a process called creaming. The relatively rare situation where the oil disperse phase tends to sink to the bottom is called downward creaming. Creaming is distinct from breaking which involves the coalescence (that is, merging together) of the droplets. Nevertheless, the emulsion droplets in the cream are in much closer proximity to one another than in the bulk of the emulsion and so droplet coalescence is easier.

Although Stokes law cannot be applied rigorously to the rate of creaming, it does give some useful indications about the factors involved. It can be written:

$$v = \frac{2gr^2(\rho_o - \rho_w)}{9\eta_w} \tag{6.1}$$

where $v$ is the sedimentation or creaming rate, $g$ is the acceleration due to gravity, $r$ the droplet radius, $\rho_o$ and $\rho_w$ are the densities of disperse phase (oil) and continuous phase (water) respectively, and $\eta_w$ is the viscosity of the continuous phase.

It is evident from Eq. (6.1), and indeed from intuition, that creaming can be minimized by increasing the density of the oil phase, for example by adding carbon tetrachloride, or by reducing the size of the droplets. In the process of homogenization the particle size is reduced to such an extent that Brownian motion is sufficient to prevent the creaming action.

## Breaking

Although the interfacial energy of the oil–water interface is reduced by the presence of an emulsifying agent, it is still positive so there is a thermodynamic drive to reduce the interfacial area. This reduction will occur if the emulsion droplets coalesce, so emulsions are thermodynamically unstable. However, even though coalescence or breaking is thermodynamically

---

**Not all creaming is bad . . .**

For some purposes it is desirable to facilitate creaming. The concentrating of rubber latex by the addition of certain gums (natural hydrophilic solids with small particle size) is an example. The original latex contains a large proportion of very small droplets which would cream very slowly, but the creaming agent causes flocculation into clusters that may contain several hundred droplets and these clusters cream at a rate appropriate to their overall size.

Another example is in the manufacture of butter, where the natural creaming process in milk (an O/W emulsion with about 4% butter fat) is accentuated by centrifugation. The separated cream (about 36% butter fat) is then agitated by churning at low temperatures to break and invert the emulsion giving butter: a W/O emulsion with about 15% water.

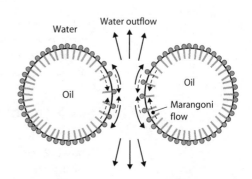

**Figure 6.1.** The approach of two oil drops forces water out from the gap between them and this drags emulsifier out as well. The interfacial tension rises in the depleted regions of the interface and generates a Marangoni flow (dashed arrows) that tends to return emulsifier and drag water back as well.

favoured, the rate at which it occurs may be very slow. By reducing the rate of creaming, as indicated above, the rate of breaking will also be reduced.

The surfaces of the emulsion droplets are coated with an adsorbed layer of the emulsifier and this can impede coalescence. There are two static effects: one is the reduction of interfacial tension caused by the emulsifier and the consequent decrease in the thermodynamic drive toward coalescence; the other is the physical barrier that the adsorbed layers impose. Dynamic effects also contribute. When two drops approach, the flow of liquid from between them (a necessary precursor to coalescence) is retarded by the concomitant dilution of the emulsifier film (Figure 6.1). This sets up a gradient in interfacial tension, higher in the interparticle region than elsewhere on the drops, so the Marangoni effect (see Section 4.10) tends to move emulsifier back to the interparticle region and to drag liquid back with it, thereby impeding coalescence. Also the viscosity of the interfacial film affects such movements (Becher, 2001, p. 136).

### Inversion

Whether an O/W or W/O emulsion is formed depends on several factors. One is the emulsifier used, another is the ratio of oil to water.

In forming an emulsion by shaking there will be initially a mixture of O/W and W/O types and the type that eventually dominates will be partly determined by the relative coalescence rates, the type with the lower coalescence rate being favoured (Davies and Rideal, 1961, p. 371ff).

In emulsifying machines, oil and water are mixed between two shearing plates. The component with the higher relative phase volume tends to form the continuous phase although the precise ratio at which the inversion from

---

Mayonnaise

A basic mayonnaise is made with egg yolk, mustard, oil, and vinegar. It is an O/W emulsion with about 60–80% oil phase. The emulsifiers in egg yolk are lecithin (10%), a good O/W emulsifier and cholesterol (1.5%), which is a good W/O emulsifier. Investigations have shown that with this ratio of lecithin to cholesterol the O/W emulsion should be unstable so that the presence of mustard is essential. The mustard is thought to act as a finely divided solid (see Section 6.3.4).

O/W to W/O varies somewhat with the HLB value of the emulsifier, the viscosity of the oil phase, and the material forming the shearing plates.

There have been a number of attempts to describe a mechanism for the inversion process but none has been generally accepted.

### 6.2.5 Emulsion rheology

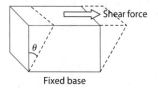

**Figure 6.2.** A shear force acting on the upper surface of the rectangular block causes a deformation that can be measured by the angle $\theta$.

Rheology is concerned with the deformation or *strain* of a material as a consequence of an applied force, the *stress*. Stress can take a variety of forms, but it is shear stress that is of interest here.

Figure 6.2 shows a shearing force, $F$, acting on the upper surface of a rectangular block of area $A$, causing the shearing strain measured by the angle $\theta$. Quantitatively, the shear stress, $\tau$, and shear strain, $\gamma$, are given by:

$$\tau = F/A \tag{6.2}$$

and

$$\gamma = \tan \theta \tag{6.3}$$

(noting that if $\theta$, in radians, is small, $\tan\theta \approx \theta$). The *shear modulus, G*, is the ratio of stress to strain:

$$G = \tau/\gamma. \tag{6.4}$$

When the shear stress is removed, a solid will tend to return to its original shape, a liquid will retain the new shape, and a plastic material may show partial recovery. The duration of the stress is also important: with a short sharp impact a liquid may behave like a solid (a skimming stone, for example, will bounce off a water surface as if the water were solid).

When a liquid is subjected to a shearing stress the concept of ideal behaviour is known as *Newtonian behaviour*. For a Newtonian liquid the shear stress is proportional to the time rate of strain ($d\gamma/dt$), and the proportionality constant is called the viscosity ($\eta$):

$$\tau = \eta \frac{d\gamma}{dt}. \tag{6.5}$$

This ideal relationship is followed by very few liquids, with water, dilute aqueous solutions and dispersions, ethanol, olive oil, and glycerol as important examples.

There is too wide a range of non-Newtonian behaviour to discuss here in detail, but two types should be mentioned as they have important practical consequences. They are shear thinning or pseudo-plasticity, where the viscosity decreases with increasing shear rate, and shear thickening or dilatancy, where the viscosity increases with increasing shear rate. There is a useful empirical equation that describes these behaviour patterns:

$$\tau = k \left( \frac{d\gamma}{dt} \right)^n \tag{6.6}$$

where $k$ is called the *consistency index* and $n$ depends on the type of flow. For shear thinning, $n < 1$; for Newtonian flow, $n = 1$; for dilatancy, $n > 1$.

---

Silly Putty

Silly Putty was a by-product of a wartime search for a rubber substitute. It is made from silicone oil and boric acid plus some minor ingredients. It provides an excellent illustration of dilatancy: it can be slowly stretched, shaped and torn, but if rolled up and thrown on the floor it will bounce like a hard ball.

---

A high viscosity for the dispersion medium of an emulsion reduces the rate of creaming and also hinders the rate of liquid outflow between two approaching droplets (Figure 6.1). Both factors reduce the rate of coalescence.

For a very dilute suspension of smooth, rigid, spherical, uncharged particles, Einstein has shown that the viscosity depends on the volume fraction of the disperse phase, $\varphi$:

$$\eta = \eta^0(1 + 2.5\varphi) \tag{6.7}$$

where $\eta^0$ is the viscosity of the dispersion medium. The shear rate, the sizes of the spheres and the range of sizes present do not affect the viscosity only the phase volume is relevant.

With emulsions the particles are, of course, not rigid and tangential stress can be transmitted from one liquid phase to the other. Thus in shear flow the dissipation of energy is less than for rigid spheres so the viscosity of a dilute emulsion is lower than that given by Eq. (6.7). If the interfacial tension is high the droplets remain spherical at low shear rates. With lower interfacial tension and/or higher shear rates, the liquid droplets may change shape to ellipsoids and even burst at high shear rates forming smaller droplets. It is also worth noting that small droplets are less easily deformed than larger droplets so the viscosity of an emulsion depends on particles size.

The charges on emulsion particles also affect the viscosity and three effects of charge on viscosity (also known as *electroviscous effects)* have been distinguished: The *primary* effect is the distortion of the electrical double layer by the shear flow (see Section 9.5); the *secondary* effect arises from the double layer repulsion between particles (see Section 9.9); and the *tertiary* effect

---

Water-based paints

These days the majority of paints are water-based, meaning that they are oil in water emulsions. The *oil* is a film-forming polymer containing a coloured *pigment* and the *vehicle* is water with some dissolved polyalcohols to assist with emulsion stability and control drying. The rheological requirements are quite complex. The dispersion must remain stable in the can, and if settling does occur redispersion must be easy; the paint must stick to the brush or roller without dripping, but spread readily over the surface being painted; the droplets must then flow sufficiently to merge and eliminate brush marks, but not form runs or sags under gravity.

Consequently the paint must be highly *shear-thinning.*

applies to poly-electrolytes and is attributed to changes in particle dimensions with changes in pH or ionic strength.

## 6.3 Emulsion stability and selection of the emulsifier

Generally, emulsions will not form without the presence of an emulsifier. The essential function of the emulsifier is to lower the interfacial tension at the oil–water interface, but there are more subtleties than this simple proposition would imply.

### 6.3.1 Early theories

The selection of a suitable emulsifier is a very important aspect of emulsion preparation and in determining whether an O/W or an W/O emulsion will be formed.

One of the first attempts to predict emulsion type is known as the Bancroft rule (Bancroft, 1913) which, in a simplified form, states that the phase in which the emulsifier is the more soluble tends to be the continuous phase.

Further understanding, based on film balance studies, came with the recognition by Langmuir and Harkins that the emulsifier forms an oriented monolayer at the oil–water interface. The oriented wedge theory of Harkins (1952) arises from this insight and suggests that the end of the emulsifier molecule with the larger size (loosely defined) will lie in the continuous phase. There are many exceptions, and the wedge concept is somewhat unrealistic as the curvature of the emulsion droplets is negligible in comparison to the size of the emulsifier molecules.

An elaboration of the Bancroft theory by Winsor (1952) considers 10 different cohesive energies, including those between the hydrophilic part of the emulsifier and water ($E_{HW}$) and between the lipophilic part of the emulsifier and oil ($E_{LO}$). When these two cohesive energies are greater than the other cohesive energies then, according to the Winsor theory, the ratio:

$$R = \frac{E_{LO}}{E_{HW}} \tag{6.8}$$

determines the nature of the emulsion: if $R < 1$, an O/W emulsion is formed; whereas $R > 1$ gives an W/O emulsion. Such emulsions are sometimes designated *Winsor I* and *Winsor II* respectively.

Schulman and Cockbain (1940) investigated mixed emulsifiers for making O/W emulsions. Their work was based on the monolayer penetration studies described in Section 5.7.2, selecting systems where the surfactant + monolayer films were stable to unusually high surface pressures. In general, the two amphiphiles in such systems have structures that enable them to pack closely together. Stable emulsions did not form when the film balance studies showed that the molecules were not able to pack into a condensed film.

There is also an appreciable contribution to stability from the visco-elastic properties of the interfacial film when the emulsifier is macro-molecular, such as gum arabic, protein, or cellulose derivatives.

If an emulsion droplet rises to the surface of the liquid, as in creaming, it could spread over the surface if the spreading coefficient is positive. Repetition of this process by other droplets will lead to the formation of a bulk phase and the breaking of the emulsion. Changes in the various interfacial tensions during this process complicate the situation, but it is clear that a negative spreading coefficient is required for emulsion stability.

## 6.3.2 Electrical aspects of emulsion stability

Many of the emulsifiers used to stabilize emulsions are ionised so the surfaces of the emulsion droplets carry an electric charge. In O/W emulsions, the presence of such surface charges will lead to the formation of an electrical double layer around the droplets. A detailed account of the electrical double layer formed at the aqueous solution–solid interface is given in Chapter 9 and reference to that material would amplify the relatively brief qualitative description that will be given here.

In essence, the charged surface attracts ions of opposite charge and repels ions of the same charge forming a diffuse layer throughout which the imbalance between oppositely charged and like-charged ions decreases with distance from the interface until a balance is reached. At the outside of this layer the electrical potential is zero as the net charge in the diffuse layer exactly balances the charge on the droplet surface. If the concentration of ions in the aqueous phase is increased the effective thickness of the diffuse layer is reduced because compensation for the surface charge occurs more readily. Refinements in the structure of this layer to allow for the size of the ions adsorbed close to the surface lead to the introduction of the Stern layer: a single layer of oppositely charged ions of finite size adjacent to the surface. The diffuse layer then extends out from the Stern layer.

When two emulsion droplets approach one another there is a repulsive interaction when their diffuse layers begin to overlap. This repulsion increases as the overlap increases and the kinetic energy of the droplets must be used to move them closer together. However, in addition to this double-layer repulsion there is a van der Waals attractive force between the droplets. This combination of double layer repulsion and van der Waals attraction generates a potential energy barrier to closer approach, and this must be overcome by the kinetic energy provided by the relative motion of the droplets (Figure 6.3). If there is sufficient kinetic energy to overcome the potential energy barrier the droplets will flocculate (but not necessarily coalesce). It is important to recognize that at a particular separation it is not the absolute value of the potential energy that determines the force experienced by the droplets, but the slope of the potential energy curve: if the potential energy increases as the droplets approach they experience a repulsive force, if it decreases they experience an attractive force.

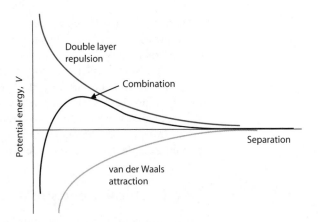

**Figure 6.3.** Schematic diagram showing the potential energy of two emulsion droplets as a function of their separation.

The height of the energy barrier can be decreased by increasing the concentration of ions in the aqueous phase and it depends also on the charge number of the added ions of opposite charge. Trivalent ions are more effective than divalent ions which are in turn more effective than singly charged ions, an effect known as the **Schulze–Hardy rule**.

The stability of O/W emulsions is therefore enhanced if the concentration of ions, particularly multivalent ions, in the aqueous phase is minimized. Conversely, flocculation can be brought about by adding salts, particularly those with multivalent ions of opposite charge to the charge on the surface of the droplets.

It is also worth noting that the addition of electrolytes can sometimes lead to inversion with a form of the Schulze–Hardy rule operating.

### 6.3.3 **Steric stabilization**

Another mechanism of stabilization is found when the emulsifier is polymeric. Here the polymer projects loops and tails into the continuous phase (Figure 6.4) and the interaction of these projections hinders the approach of emulsion droplets to one another. An electric charge with its associated double layer is not necessary: even uncharged droplets can be stabilized by this mechanism. Thus steric stabilization may be effective in non-aqueous dispersion media where charge stabilization is unsatisfactory. In many cases stabilization is due to both steric and electrostatic effects and this known as *electrosteric* stabilization. The topic has been reviewed in detail by Vincent (1974) and by Napper (1983).

Often the polymers are block or graft copolymers with hydrophilic segments and lipophilic segments so that the polymer is adsorbed with its hydrophilic segments in the aqueous phase and its lipophilic segments in the oil phase. Strong attachment of the polymer to the disperse phase is essential, but it is obviously the conformation of those parts of the polymer that are in the dispersion medium that are relevant to the stability of the emulsion.

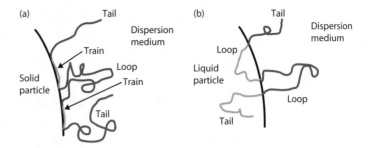

**Figure 6.4.** Schematic diagram of a polymer molecule at a surface of (a) a solid particle and (b) a liquid particle.

Factors that need to be considered when two sterically stabilized emulsion droplets approach one another include:

- loss of configurational entropy due to the fact that the droplets are excluded from the volume occupied by the adsorbed layer on the other droplet: termed the *volume restriction* effect;
- in regions where the adsorbed layers on the two particles interpenetrate there is an increase in *osmotic pressure* which tends to draw more solvent into that region tending to force the droplets apart;
- there may also be changes in the conformations of the polymer loops and tails as the two droplets approach.

Steric stabilization is used extensively in the pharmaceutical and cosmetic industries, with natural polymers predominating.

---

### Casein particles in milk

The name *casein micelles*, often applied to these particles, is unfortunate as it implies a similarity to surfactant micelles that does not exist. Surfactant micelles are in dynamic equilibrium with the surrounding liquid and surfactant molecules are constantly entering and leaving the micelles. On the other hand, the particles known as *casein micelles* are aggregates an order of magnitude larger than normal surfactant micelles, formed irreversibly from various milk proteins with calcium phosphate and some citrate as a binder. They are responsible for the milky appearance of skim milk.

The casein particles in milk consist of a mixture of smaller casein particles of various types: $\alpha$-, $\beta$-, and $\kappa$-casein. The particles have a disordered open structure that is strongly hydrated and sufficiently porous to allow other proteins, such as enzymes, to penetrate into them. They contain most of the calcium that is present in milk. The $\kappa$-casein is hydrophilic and mostly occurs at the surface of the casein particles where it is thought to stabilize the suspension by the steric mechanism and through double layer repulsion arising from the charges on the $\kappa$-casein tails. There may also be a small contribution to stability from the negative charges on the particles themselves. Enzymatic removal of the $\kappa$-casein by rennin causes coagulation. This is the basis of cheese manufacture.

In certain circumstances the addition of polymer can lead to flocculation. Generally in such cases, the polymer has a high molar mass and is either non-ionic or carries a charge of the same sign as the disperse particles. Bridging is the usual mechanism, where the polymer is attached to both particles.

### 6.3.4 Stabilization by solid particles

A brief mention should be made of the fact that emulsions may be stabilized by finely divided solid particles. This phenomenon is not used commercially to prepare emulsions but has a certain nuisance value in that it forms and stabilizes emulsions in unwanted situations.

Particles that stabilize emulsions need to be amphiphilic so that they sit at the oil–water interface. However, it seems that the presence of such particles does not make emulsification easier: it is only after the emulsion has formed that they migrate to the interface and improve the stability.

### 6.3.5 Emulsifiers and the HLB system

At present the best guide for the selection of an emulsifier is provided by a classification according to their hydrophilic-lipophilic balance or HLB value. At first, HLB values were determined empirically from the emulsifying and solubilizing properties of various surfactants and this is still the preferred method. Attempts have been made to calculate the HLB value using the chemical formula and empirically determined contributions from groups present in the molecule but the results have not always been successful, nor have efforts to develop theoretical approaches. Some reasonable HLB values have been reported from a linear relationship between the HLB of the emulsifier and the logarithm of the ratio of its solubilities in water and in oil (often toluene) (Becher, 2001).

For a W/O emulsion the HLB value should be between 3.5 and 6, while for a O/W emulsion a value between 8 and 18 is desirable. Many of the common surfactants lie outside these ranges (sodium lauryl sulfate, for example has a value of 40, while for oleic acid the value is 1) and thus have little use as emulsifiers, but there are numerous other surfactants that are suitable. Some of these are listed in Table 6.2 and a more extensive list is given by Becher (2001).

It is important to emphasize that emulsion type is not solely determined by the HLB value of the emulsifier. Factors such as the volume ratio, the type of oil, electrolytes in the aqueous phase, and temperature also have an influence.

Raising the temperature of a solution of a non-ionic surfactant decreases its hydrophilicity and consequently decreases the HLB value. For an O/W emulsion stabilized by such an emulsifier, raising the temperature will destabilize the emulsion and may eventually lead to inversion: the formation of a W/O emulsion. This is sometimes called the *phase inversion temperature* or *PIT*. There is a good correlation between the PIT and the cloud point of the non-ionic emulsifier (see Section 4.7.2).

Many natural emulsions are stabilized by macromolecules such as proteins (milk, for example) or polysaccharides such as gums and starches. Emulsion

**Table 6.2.** HLB values for some amphiphiles.

| Amphiphile | HLB | Amphiphile | HLB |
|---|---|---|---|
| Sodium lauryl sulfate | 40 | POE-2 oleoyl alcohol | 4.9 |
| Ammonium lauryl sulfate | 31 | Sorbitan monostearate | 4.7 |
| Sodium oleate | 18 | Span 80 (sorbitan monooleate) | 4.3 |
| Tween 80 (sorbitan monooleate EO$_{20}$) | 15 | Glycerol monostearate | 3.8 |
| Tween 81 (sorbitan monooleate EO$_6$) | 10 | Glycerol monooleate | 3.4 |
| | | Sucrose distearate | 3.0 |
| Calcium dodecylbenzene sulfonate | 9.0 | Sorbitan tristearate | 2.1 |
| Sorbitan monolaurate | 8.6 | Glyceryl dioleate | 1.8 |
| Soya lecithin | 8.0 | Cetyl alcohol | 1.0 |
| Sorbitan monopalmitate | 6.7 | Oleic acid | 1.0 |
| Glycerol monolaurate | 5.2 | | |

stabilization with synthetic polymers is also possible. Emulsions stabilized by such materials can be very stable and consideration of the processes described in Figure 6.1 suggests an explanation.

It is also worth mentioning that the ionized emulsifiers in Table 6.2 all have high HLB values. This can be attributed to the strong interaction with water molecules exerted by the charged head groups. Within this group, the relatively low value shown for the oleate chain compared with lauryl chains also reflects the presence of the stiff double bond in the oleate chain hindering its ability to pack with other oleate chains, whereas the more flexible, fully saturated, lauryl chains can pack together easily.

## 6.4 Microemulsions

There is some debate as to whether these systems should properly be described as emulsions or as swollen micelles. What is clear is that there is a continuum of droplet sizes ranging from small swollen micelles with diameters ranging upwards from 1 nm to microemulsions with diameters up to 60 nm. The question is where the boundary should be placed. This situation is further complicated by the possible formation in certain cases of a bicontinuous phase (see Figure 4.18) (Evans *et al.*, 1986) that may not always be recognized as such.

There is a definite optical distinction between microemulsions and normal emulsions. In microemulsions the droplets are too small to scatter light so the liquids are optically clear, whereas normal emulsions scatter light strongly and appear white and opaque (like milk). There is also another, very important, distinction: normal emulsions are thermodynamically unstable, as we have seen, whereas microemulsions are thermodynamically stable. There is thus a sense that the microemulsions are lyophilic colloids whereas normal emulsions are lyophobic (see Section 9.7.1).

Furthermore there are also clear distinctions between microemulsions and micelles. As material to be solubilized is added to a micellar system, the originally isotropic (approximately spherical) micelles can swell only to a limited extent before they must reorganize to different, anisotropic, shapes (cylinders or lamellae) in order to accommodate the extra material (see Figure 4.18). In contrast, microemulsion droplets are able to accept significantly larger amounts of the disperse phase while retaining their isotropic nature, although it is essential that extra emulsifier should be available to maintain the interface during this swelling process. Furthermore, in a micellar system there is chemical equilibrium throughout the entire system: continuous phase, disperse phase, and interface. This is not necessarily the case with microemulsions (Hunter, 1989).

As mentioned earlier, the discovery of microemulsions arose from the work of Schulman and associates on the penetration of insoluble monolayers by water soluble surfactants (see Section 5.7.2). They observed that in some systems the penetrated monolayer appeared to be stable up to very high surface pressures and claimed that in certain cases a negative surface tension was reached. As a negative surface tension would lead to a spontaneous increase in the interfacial area, further investigations of such systems were begun and led to the discovery of spontaneous emulsification and microemulsions.

An increase in the interfacial area would sooner or later lead to a dilution of the surface layer of emulsifier so that, normally, a negative value of the interfacial tension would only be temporary. Indeed, it must be temporary. If the interfacial tension remained negative the interface would expand until molecular dispersion was reached: the material intended for the disperse phase would have dissolved.

Thus one method for forming microemulsions has certain parallels with the penetration of monolayers, suggesting that two surfactants might be required and that one should be water soluble (high HLB) and the other oil soluble (low HLB). These are sometimes referred to as the *surfactant* and the *cosurfactant*, respectively. Fairly large amounts, 15–25 wt.%, of surfactant plus cosurfactant are usually needed. Microemulsions can also be formed using a single surfactant such as certain double chain ionic surfactants (such as didodecyl-dimethyl-ammonium bromide) or non-ionic surfactants. It is probably significant that these double chain surfactants are practically insoluble in both water and oil. In all of these systems the proper balance of hydrophilic and lipophilic tendencies is essential.

Because of the small size of the microemulsion droplets their surfaces are highly curved and it is therefore possible to predict, to some extent, whether a particular surfactant (and cosurfactant) will promote an O/W or a W/O emulsion by examining the shape of the molecules. Relatively large hydrophobic segments will favour W/O emulsions, while relatively large hydrophilic groups will favour O/W emulsions. This cone-like approach has been discussed at length by Israelachvili (1991, p. 380).

For W/O microemulsions, penetration of molecules of the oil between the hydrocarbon chains of the interfacial film can be a significant factor (dependent on the chain length of the oil) and tends to increase the bulk of the lipophilic layer and decrease the size of the water droplets.

The interactions between microemulsion droplets are similar to those between droplets in a macroemulsion, but are much less significant with regard to emulsion stability. Thus, for example, while double layer interaction and steric effects may be present in a microemulsion they do not provide the main contribution to stability.

Microemulsions are finding increasing applications in foods, pesticides, and polishes. There has also been intense activity in tertiary oil recovery. The first (primary) recovery of oil from a well is followed by pumping water into the oil reservoir to displace further oil (secondary recovery). About a third of the oil still remains and so a concentrated surfactant solution is injected to markedly reduce the oil/water interfacial tension. An O/W microemulsion forms and, pushed along by a viscous polymer solution, drives the bulk of the remaining oil upwards for recovery.

The topic has been reviewed by Shinoda and Friberg (1975).

## 6.5 Emulsion polymerization

Polymerization reactions are exothermic so the heat generated in a bulk phase polymerization can often cause problems such as inadequate control of the process and variable quality. If, on the other hand, the reaction is carried out in a dispersed phase in water, the heat is readily removed and the process can be well controlled.

The monomer is prepared as an emulsion in water stabilised by a surfactant. The initiator is dissolved in the water and provides a source of free radicals by reacting with the small number of monomer molecules in the aqueous phase.

$$I^{\bullet} + M \longrightarrow M^{\bullet}.$$

After this *initiation step* the monomer radical reacts, in the *propagation step*, with other monomer molecules in the aqueous phase and in a sequence of similar reactions generates an *oligomer*:

$$M^{\bullet} + M \longrightarrow MM^{\bullet}$$
$$MM^{\bullet} + M \longrightarrow MMM^{\bullet}$$

and so on. The oligomers, which are minute polymer particles, are solubilized in the surfactant micelles where they continue to propagate by reacting with monomer molecules that diffuse into the micelles. The reaction eventually stops, in the *termination step*, when two radicals meet and their unpaired electrons (the source of the radical property) become paired:

$$\ldots MMM^{\bullet} + \ldots MM^{\bullet} \rightarrow \ldots MMMMM \ldots$$

The emulsified monomer is the reservoir for monomer in the aqueous phase. The polymerization reaction does not occur directly inside the emulsion droplets. These colloidal dispersions are usually called *latices* because of their resemblance to rubber latex.

By careful control of the conditions it is possible to prepare monodisperse polymer colloids. Furthermore, the particles are perfectly spherical. It is also possible to incorporate ionizable groups into the particle surfaces and thus to control the particle charge by controlling the pH of the dispersion medium.

Most polymers are now produced by emulsion polymerization.

---

### Monodisperse colloids from emulsion polymerization

Spherical colloidal particles can be prepared by emulsion polymerization. The procedure used by Ottewill and Shaw (1967) is described below. It relies on the observation that nuclei for the polymerization only form in the micelles of the emulsifier so there are two stages in the procedure: formation of seed nuclei with the emulsifier concentration well above the cmc; and growth on these nuclei with the emulsifier concentration below the cmc.

A seed dispersion of polystyrene is prepared by ultrasonic agitation of a mixture of styrene, sodium dodecanoate, and water (at pH 10) to form an emulsion. This emulsion and hydrogen peroxide as initiator are then stirred at 70 °C until the polymerization reaction is complete (about 6 h) giving nuclei of about 50 nm diameter. Excess sodium dodecanoate is then removed by dialysis against sodium hydroxide solution (pH 10). For the growth stage more styrene, water, and a limited amount of sodium dodecanoate are added, the mixture is emulsified, and polymerization initiated with peroxide as before. The resulting latex is purified by filtration and by dialysis.

The growth stage can be repeated a number of times to produce the required particle size. The size distribution becomes narrower as the particle size increases as the smaller particles grow at a faster rate than the larger ones. Initiation by hydrogen peroxide avoids the incorporation of sulfate groups into the colloid which is a feature of colloids formed with potassium persulfate as initiator.

---

### Nylon rope experiment

Reaction at a liquid–liquid interface is clearly seen in the formation of 'nylon rope' in a laboratory demonstration. A solution of 8 g sodium carbonate and 4.5 g of hexamethylenediamine in 100 cm$^3$ water is carefully poured on top of a solution of 3 cm$^3$ of sebacoyl chloride in 100 cm$^3$ of perchloroethylene. A visible film forms at the interface and this may be grasped with tongs and drawn up out of the beaker as a 'rope'. Using a steady rate of withdrawal, a considerable length of nylon rope can be produced.

## 6.6 Liquid–liquid extraction

The transfer of a solute from one liquid phase to another liquid phase is a topic of considerable industrial importance and there are a number of engineering texts devoted to it (e.g. Sherwood *et al.*, 1975).

It is convenient to discuss transport processes in terms of a *flux* (rate of movement of solute through unit area) and *resistances* to movement. The entire transport pathway may involve several resistances and, just like electrical resistances, these may be in parallel but more usually are in series. Normally the driving force for movement is expressed as a difference in solute concentration with the equilibrium situation as reference point. At equilibrium the flux goes to zero and hence the driving force must also be zero. Thus we have a relationship between these quantities that is analogous to Ohm's law in electrical theory:

$$J \equiv \frac{1}{A}\frac{dn}{dt} = \frac{\Delta c}{\sum r} \tag{6.9}$$

where $J$ is the flux (defined by the identity), $\Delta c$ is the driving force, and $\sum r$ is the sum of the resistances in the transport pathway.

As our present emphasis is on events at interfaces it is convenient to assume that the bulk liquids are well mixed, possibly turbulent, with uniform concentrations of solute. In these circumstances the resistances to transport of solute are confined to the interfacial region and broadly can be attributed to liquid phase $\alpha$, the interface $\sigma$ itself, and liquid phase $\beta$. The total resistance is therefore:

$$r = r^{\alpha} + r^{\sigma} + r^{\beta} \tag{6.10}$$

where the resistances assigned to phases $\alpha$ and $\beta$ refer to thin layers adjacent to the interface. This is a form of the two-film model introduced by Whitman (1923).

The resistance of the interface may be affected by the presence of surfactant (see particularly Section 4.9.3 and Section 5.9.1). Furthermore the transfer of solute across the interface may cause turbulence which affects the resistances and in some circumstances may lead to emulsification.

## 6.7 Membranes

It is convenient to discuss here the situation where the two liquid phases are miscible, often with the same solvent, but mixing is prevented by a membrane at the interface. The subject has a long history as Graham in the mid-nineteenth century, following earlier work, used membranes to purify colloids by dialysis (see Section 9.7.2). The membranes used for dialysis and other laboratory procedures were initially of natural origin, such as pig and fish bladders; parchment, collodion, and cellophane, while the more recent introduction of polymers now provides a wide variety of membrane types. Natural membranes are essential components in all biological systems: they form the containers that retain the contents of animal and plant cells, and control the movement of various molecules, from oxygen and carbon dioxide to food and waste materials, from one part of the organism to another. They are discussed at greater length in Section 10.5.

The essential feature of both artificial and natural membranes is their selective permeability. Their function is not simply to separate two bulk liquid phases, but to permit the transport of selected materials from one liquid to the other while preventing the passage of others. In early work on colloids the term *semipermeable membrane* was used to describe membranes that would allow the passage of simple inorganic ions and solvent molecules while preventing the larger particles of colloidal dispersions from passing through.

The present discussion will be divided into three parts, dealing with *artificial membranes, black lipid membranes*, and with *self-assembled membranes*. By artificial membranes we mean membranes that have been introduced artificially to the system, even though they may have been formed elsewhere by some natural process: pig bladder, for instance. In contrast, the self-assembled membranes form spontaneously from components of the system, often amphiphiles. Black lipid membranes are essentially artificial membranes, but they are so different from other artificial membranes that they need to be treated separately.

### 6.7.1 Artificial membranes

#### Osmotic pressure

The ability of a semipermeable membrane to exhibit selectivity in the passage of molecules is clearly demonstrated by the phenomenon of osmosis. If, for example, an aqueous solution of sucrose is placed in an inverted thistle funnel with the wide end tightly covered by cellophane, and this end is immersed in water, water will flow through the cellophane, but sucrose is unable to pass through, so the liquid level in the tubular part of the funnel rises (Figure 6.5). This process will continue until equilibrium is reached.

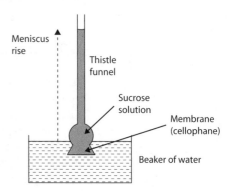

**Figure 6.5.** Apparatus for demonstrating osmotic flow.

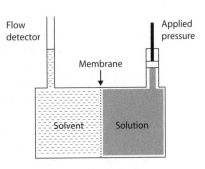

**Figure 6.6.** Schematic diagram of apparatus for the measurement of osmotic pressure. The osmotic pressure is the pressure that must be exerted on the solution that is just sufficient to prevent movement of solvent through the membrane into the solution.

At equilibrium, the pressure exerted by the movement of solvent into the funnel is exactly balanced by the hydrostatic pressure from the column of liquid in the funnel. This pressure is the osmotic pressure. The apparatus shown in Figure 6.5, while useful as a demonstration of osmosis, is not satisfactory for measuring osmotic pressure as the concentration of the solution in the funnel changes as the osmotic flow proceeds and the solute concentration is probably not uniform.

Osmotic pressure is defined as the pressure just required to prevent osmotic flow, and in its measurement the extent of osmotic flow must be minimized. A suitable design is shown schematically in Figure 6.6.

The analysis of this experiment is treated in most physical chemistry texts so will be only briefly outlined here. At equilibrium the change in free energy caused by the presence of solute must be balanced by the change coming from the increased pressure (the osmotic pressure, $\Pi^{os}$).

For the effect of solute (B) on the vapour pressure of solvent (A),

$$\Delta G = n_A RT \ln (p/p^\bullet)$$

and for the pressure increase,

$$\Delta G = V \Pi^{os}$$

so that at equilibrium,

$$n_A RT \ln (p/p^\bullet) + V \Pi^{os} = 0. \tag{6.11}$$

If the solution is dilute we can assume Raoult's law for the solvent,

$$p/p^\bullet = x_A = 1 - x_B$$

and use the approximation

$$\ln (1 - x_B) \approx -x_B$$

giving

$$\Pi^{os} V / n_A = RT x_B \tag{6.12}$$

where $V/n_A$ is the molar volume of *solvent*. With some further approximations Eq. (6.12) can be changed into the empirical equation devised by van't Hoff in 1885:

$$\Pi^{os} V' = n_B RT \tag{6.13}$$

where $V'$ is the volume of solution rather than solvent. This equation provided a reasonable fit to much of the experimental data available to that time, but there was one important anomaly: when the solute was ionised a numerical factor, the van't Hoff $i$ factor, had to be inserted. It is a matter of historical record that these $i$ factors provided significant experimental support to theories of ionization developed at about 1887 by Arrhenius and later workers.

The requirement to insert the van't Hoff factor into equations for osmotic pressure is a consequence of the colligative nature of osmotic pressure: with colligative properties the value depends on the total concentration of solute particles and not on the nature of these particles. Thus when, for example, surfactant molecules aggregate to form micelles the dependence of osmotic pressure on total surfactant concentration changes markedly, as shown in Figure 4.15.

### Dialysis

Semipermeable membranes are frequently used to purify colloidal dispersions, sols or emulsions. Here the membrane must be able to allow small ions to pass through as well as solvent molecules while retaining the colloid particles. Cellophane (a reprecipitated cellulose) is commonly used. For purification purposes the cellophane comes as a tube that can be cut and tied to form a bag. Boiling in several changes of distilled water to remove soluble impurities (introduced to prevent embrittlement) is usually desirable before use. The impure colloid is sealed in the bag and immersed in distilled water which is changed several times over a period of days. Ions present in the colloid as impurities will diffuse out of the bag, driven by the concentration difference, while water will tend to enter the bag by osmosis. It is therefore important that the bag should be strong enough to withstand the osmotic pressure that will develop.

### Donnan membrane equilibrium

If a dispersion of large charged molecules such as a protein is placed on one side of a semipermeable membrane and solvent or a solution of a simple electrolyte on the other, there will be a redistribution of ions. In the situation illustrated in Figure 6.7, the large $P^-$ ions are unable to pass through the membrane, but the $Na^+$ and $Cl^-$ ions can. We end up with an asymmetric distribution of the small ions, the *Donnan effect*, and an electrical potential across the membrane.

Initially there are no $Cl^-$ ions in solution $\alpha$, so there is a net movement of $Cl^-$ from $\beta$. To preserve electrical neutrality, this movement must be accompanied by the movement of an equal number of $Na^+$ ions from $\beta$ to $\alpha$. This process continues until the activity products, $a(Na^+) \cdot a(Cl^-)$, are the same on both sides of the membrane. This generates a difference in electrical potential across the membrane: the Donnan potential. Further treatment of this system requires an understanding of electrochemical potentials ($\overline{\mu}_i$) which are described below.

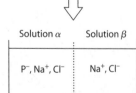

| Solution $\alpha$ | Solution $\beta$ |
|---|---|
| $P^-$, $Na^+$ | $Na^+$, $Cl^-$ |

Initial situation

| Solution $\alpha$ | Solution $\beta$ |
|---|---|
| $P^-$, $Na^+$, $Cl^-$ | $Na^+$, $Cl^-$ |

Equilibrium situation

**Figure 6.7.** The movement of ions in the Donnan membrane equilibration process.

---

### Electrochemical potential

In dealing with the movement of ions as distinct from neutral molecules, it is necessary to modify the chemical potentials of the ions to take into account the electrostatic free energy. Thus the electrochemical potential is defined as

$$\bar{\mu}_i = \mu_i + z_i F \Phi \tag{6.14}$$

where $z_i$ is the charge number of the $i$ ion, $F$ is the Faraday constant, and $\Phi$ is the electrostatic potential. Assuming that we can use concentrations rather than activities in the expression for free energy, we have for the movement of $i$ from phase $\alpha$ to phase $\beta$:

$$\bar{\mu}_i^\beta - \bar{\mu}_i^\alpha = \mu_i^\beta - \mu_i^\alpha + z_i F(\Phi^\beta - \Phi^\alpha)$$

$$= RT \ln\left(\frac{c_i^\beta}{c_i^\alpha}\right) + z_i F(\Phi^\beta - \Phi^\alpha). \tag{6.15}$$

At equilibrium the electrochemical potentials must be equal.

---

At the Donnan equilibrium

$$\bar{\mu}_{Na}^\beta - \bar{\mu}_{Na}^\alpha = 0 = \bar{\mu}_{Cl}^\beta - \bar{\mu}_{Cl}^\alpha$$

and the potential difference across the membrane must be the same for both the $Na^+$ and $Cl^-$ ions, so that:

$$\Phi^\beta - \Phi^\alpha = -\frac{RT}{z_{Na}F} \ln\left(\frac{c_{Na}^\beta}{c_{Na}^\alpha}\right) = -\frac{RT}{z_{Cl}F} \ln\left(\frac{c_{Cl}^\beta}{c_{Cl}^\alpha}\right). \tag{6.16}$$

Furthermore, as $z_{Na} = -z_{Cl}$

$$\frac{c_{Na}^\beta}{c_{Na}^\alpha} = \frac{c_{Cl}^\alpha}{c_{Cl}^\beta}$$

whence

$$c_{Na}^\beta c_{Cl}^\beta = c_{Na}^\alpha c_{Cl}^\alpha. \tag{6.17}$$

This relationship, with activities replaced by concentrations, confirms the statement made earlier about the equilibrium situation.

The condition of electrical neutrality for each solution requires that

$$c_{Na}^\beta = c_{Cl}^\beta = c^\beta$$

and

$$c_{Na}^\alpha = c_{Cl}^\alpha - z_p c_p^\alpha$$

(noting that $z_P$ is negative). Substituting these two equations into (6.17) gives a quadratic which yields the approximate solution:

$$\left(\frac{c_{Na}^\alpha}{c_{Na}^\beta}\right) = 1 - \frac{z_p c_p^\alpha}{2 c_{Na}^\beta}. \tag{6.18}$$

From this relationship and (6.16) the Donnan potential $(\Phi^\beta - \Phi^\alpha)$ can be calculated from the characteristics of the polymer and the concentration of $Na^+$ in the external solution.

Generally there would be a tendency for water to flow by osmosis from solution $\beta$ to solution $\alpha$. The osmotic pressure applied to solution $\alpha$ to prevent this flow would contribute an additional term to the definition of electrochemical potential, but this term is usually small and has been neglected in the above treatment.

### 6.7.2 Black lipid membranes (BLMs)

A wide range of amphiphiles may be used to form black lipid membranes between two aqueous solutions. The two amphiphile monolayers are arranged with their hydrophilic groups facing out towards the aqueous solutions, and their hydrophobic portions towards one another, usually with an oil layer between them. Typically the area of a BLM is quite small ( <1 mm diameter).

A BLM may be formed by placing a drop of solution containing the amphiphile in a small hole in a partition separating two aqueous solutions. Another method is a variation of the Langmuir–Blodgett technique for depositing monolayers on solid substrates. In this case the substrate has a small hole in it and the monolayer bridges over the hole on both sides as the substrate is immersed. After initial formation a BLM formed with solvent will begin to thin (draining up or down depending on the relative densities of solvent and aqueous solution) and eventually show interference colours and ultimately black spots which merge to cover the entire film.

Fundamental to most studies involving a BLM is knowledge of its thickness. This is usually determined by an optical reflectance technique with the data modelled by a single- or a triple-layer (two surfactant layers and the solvent core) model.

One of the main objectives in studying BLMs is to mimic biological membranes. This usually requires the addition of non-lipid materials either during BLM formation or after formation. Transport and electrical phenomena are of major biological interest and will be discussed further in Chapter 10.

For a review of this topic see H.Ti. Tien (1988).

### 6.7.3 Self-assembled membranes

We have already seen that amphiphiles in solution can associate to form micelles, but other structures are also possible. Micelles can be spherical, tubular or sausage shaped, lamellar, or inverted. Bilayer structures can be simple sheets, or curved into more or less spherical closed structures known as liposomes or vesicles. Furthermore, there can be several stacked bilayers in these various configurations. Finally, bicontinuous structures can be formed. Generally within these various structures the hydrophobic parts of the amphiphiles are fluid-like, rather than solid, with vigorous thermal motion within the aggregate.

The forces involved in forming a particular aggregate comprise the attractive/repulsive forces between amphiphile molecules, the interaction energy, and the ability of the molecules to pack together. The solvent will usually contribute to these factors.

In an aqueous environment the hydrophobic segments of the amphiphiles will tend to attract one another whereas the head groups will be attracted to water molecules and may also be charged, both factors leading to repulsion. Combination of these attractive and repulsive forces leads to an energy minimum at an area per molecule, $\hat{A}_o$, called the *optimal surface area per molecule*.

The nature of the structures that form when amphiphiles assemble together is, according to Israelachvili (1991, p. 370), determined to a large extent by their packing parameter or shape factor:

$$\phi = v/\hat{A}_o l_c \tag{6.19}$$

where $v$ is the volume of the hydrophobic part of the amphiphile, assumed to be fluid and incompressible, $\hat{A}_o$ is the optimal area, and $l_c$ is the critical chain length (a semi-empirical parameter somewhat shorter than the length of the fully extended chains).

For a given molecule in a given solvent both the minimum interaction energy and the packing parameter will be fixed, but there is not usually one unique shape that satisfies these requirements. Several structures are possible so the determining factor will be the entropy change of the system on aggregation, and this will normally mean that the smallest possible structure consistent with the energy and structural parameters of the amphiphile will be favoured.

Israelachvili discusses the various shapes of amphiphile aggregates and shows how they depend mainly on the packing parameter. His classification is summarized in the following.

- Spherical micelles are formed by molecules with large head groups and single alkyl chains giving essentially a conical shape with $\phi < 0.3$.

- Cylindrical micelles require molecules with smaller head groups, giving a truncated cone shape and $0.3 < \phi < 0.5$.

- Curved bilayers are formed by molecules with large head groups and large tail volumes (often double chained) giving a truncated, almost cylindrical, cone shape with $0.5 < \phi < 1$. Structures in this category include the liposomes and vesicles.

- Planar bilayers are formed when the areas of the head group and of the tail section are about equal, giving a cylindrical shape and $\phi \approx 1$.

- Inverted micelles arise when the head group has a smaller area than the tail region giving an inverted truncated cone shape and $\phi > 1$.

Changes in the dispersion medium, such as the addition of electrolytes, can change the packing parameter, principally through changes to $\hat{A}_o$, and this can lead to a change from one aggregate structure to another. Of course, relatively small changes in the amphiphile itself, such as the introduction of double bonds or small branches, will increase $v$ and may also cause a structure change. Temperature also has an effect, but because it can alter both $\hat{A}_o$ and $l_c$ the situation is more complex.

## SUMMARY

The liquid–liquid interface concerns emulsions, liquid–liquid extraction, and membranes.

An **emulsion** is a dispersion of small droplets of the *disperse phase* in a liquid *dispersion medium*. As these two liquids must be mutually immiscible we can designate one as an *oil* and one usually as *water*. There are thus two types of emulsion: *oil-in-water*, O/W, and *water-in-oil*, W/O. The emulsion must also contain an emulsifier to reduce its thermodynamic instability and stabilize it. Emulsifiers are usually amphiphiles, and they can be graded according to their hydrophilic-lipophilic balance (HLB). Methods of preparation and the properties of emulsions are described with particular emphasis on stability and rheology. Microemulsions are thermodynamically stable and have smaller particles than normal emulsions.

**Liquid–liquid extraction** is described in terms of resistances to the transfer of a solute from one liquid phase to another.

**Membranes** can be classified as *artificial, black-lipid,* or *self-assembled*. The function of a membrane can be described by the term *selective permeability*: the ability to allow selected substances to pass through while preventing the passage of others. This leads to the concepts of *osmotic pressure* and *Donnan membrane equilibrium*. In discussing self-assembled membranes the relevance of the *shape factor* of the amphiphile in promoting various structural forms within a fluid medium is considered.

## FURTHER READING

Becher, P. (2001). *Emulsions: Theory and Practice, 3rd edn.* Oxford University Press, New York.

Davies, J. T. and Rideal, E. K. (1961). *Interfacial Phenomena.* Academic Press, New York.

Dickinson, E. (1992). *An Introduction to Food Colloids.* Oxford University Press, Oxford. Also other similar titles by this author.

Hunter, R. J. (1993). *Introduction to Modern Colloid Science.* Oxford University Press, Oxford.

Hunter, R. J. (2001). *Foundations of Colloid Science.* Clarendon Press, Oxford.

Israelachvili, J. N. (1991). *Intermolecular and Surface Forces, 2nd edn.* Academic Press, London.

Napper, D. H. (1983). *Polymeric Stabilization of Colloidal Dispersions.* Academic Press, New York.

Sherwood, T. K., Pigford, R. L., and Wilke, C. R. (1975) *Mass Transfer.* McGraw-Hill Kogakusha, Tokyo.

Tien, H. Ti. *in* I. B. Ivanov (ed.) (1988). *Thin Liquid Films: Fundamentals and Applications.* Marcel Dekker, New York.

## REFERENCES

Bancroft, W. D. (1913). *J. Phys. Chem.* **17**, 501.

Bancroft, W. D. (1915). *J. Phys. Chem.* **19**, 275.

Evans, D. F., Mitchell, D. J., and Ninham, B. W. (1986). *J. Phys. Chem.* **90**, 2817.

Girifalco, L. A. and Good, R. J. (1957). *J. Phys. Chem.* **61**, 904.

Harkins, W. D. (1952). *The Physical Chemistry of Surface Films*, Reinhold, New York, p. 83.

Napper, D. H. (1983). *Polymeric Stabilization of Colloidal Dispersions*, Academic Press, New York.

Ottewill, R. H. and Shaw, J. N. (1967). *Kolloid Z. u. Z. Polymere* **265**, 161.

Schulman, J. H. and Bowcott, J. E. (1955). *Z. Elektrochem.* **59**, 283.

Schulman, J. H. and Cockbain, E. G. (1940). *Trans Faraday Soc.* **36**, 651.

Schulman, J. H. and Montagne, J. B. (1961). *Ann. New York Acad. Sci.* **92**, 366.

Shinoda, K. and Friberg, S. (1975). *Adv. Colloid Interface Sci.* **4**, 281.

Vincent, B. (1974). *Adv. Colloid Interface Sci.* **4**, 193.

Wachtel, R. E. and LaMer, V. K. (1959). *J. Phys. Chem.* **63**, 768.

Wachtel, R. E. and LaMer, V. K. (1962). *J. Colloid. Sci.* **17**, 531.

Whitman, W. G. (1923). *Chem. Met. Eng.* **29**, 146.

## EXERCISES

**6.1.** A spherical oil drop of radius 1 cm is dispersed as an emulsion in water. The oil particles in the final emulsion have a radius of 0.1 µm. Calculate the energy change for the dispersion process assuming that the interfacial tension remains constant throughout at (a) $30\,\text{mN}\,\text{m}^{-1}$ and (b) $5\,\text{mN}\,\text{m}^{-1}$.

**6.2.** Calculate the osmotic pressures of aqueous solutions of sucrose and sodium chloride at molal concentrations of $0.8\,\text{mol}\,\text{kg}^{-1}$. The temperature both solutions is 298 K and the density of water at this temperature is $1.00\,\text{g}\,\text{cm}^{-3}$. Repeat the calculations using the more approximate equation of van't Hoff.

**6.3.** Calculate the free energy changes for the two dissolution processes in Exercise 6.2.

**6.4.** After equilibration the Donnan potential across a membrane separating a solution of a negatively charged polyelectrolyte and a solution of sodium chloride is found to be 45 mV (with the polyelectrolyte solution at the lower potential). Calculate the ratio of sodium ion concentrations in the two solutions. (Ignore any osmotic movement.)

**6.5.** If, in Exercise 6.4, the concentration of polyelectrolyte is $0.2\,\text{mol}\,\text{dm}^{-3}$ and the sodium ion concentration in the other solution is $3\,\text{mol}\,\text{dm}^{-3}$, what is the charge number of the polyelectrolyte?

**6.6.** Derive Eq. (6.13) from (6.12) and state the assumptions made.

# 7 The surfaces of solids

## 7.1 Introduction

Although the surface of polished metal, for example, might appear smooth and clean from a macroscopic point of view, the picture is very different when we observe it at a microscopic level. The outer layers of any surface exposed to air even for a very short time contain a variety of contaminants (shown schematically in Figure 7.1), so that the top layers are generally of very different composition to the bulk. In order to understand more about what a solid surface is really like at the molecular and atomic level, we think of a clean surface, which for crystalline materials such as metals and semiconductors can be considered to be the truncation of the bulk structure of a perfect crystal.[5] When imaged by techniques that can achieve resolution at a molecular level, such as scanning probe microscopy methods described later in this chapter, it is apparent that between smooth areas (called terraces) are a variety of defects, which can be defined as steps, kinks, adatoms, and vacancies (Figure 7.2).

It is important to realize that the local distribution of atoms around an individual atom varies according to its exact location on the surface, even on the surface of a perfect crystal, and correspondingly the electronic properties of the atoms are not uniform. This is a significant consideration when studying surfaces both experimentally and theoretically. Any information

**Figure 7.1.** Cross-section view of a solid surface of a pure metal exposed to air. The metal atoms are shown in blue, while contaminants are black and white circles.

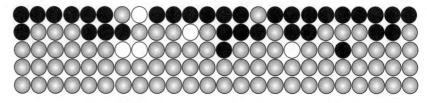

---

5. The surfaces of amorphous materials such as polymers need to be treated differently.

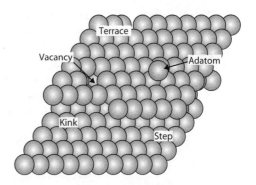

**Figure 7.2.** Hard sphere representation of surface features of a single crystal at high magnification.

gained from surface-sensitive methods is an average of many different types of surface site, and in real samples with impurities, different compositions as well. In this chapter we will concentrate on the nature and study of surfaces of pure metals and semiconductors. In Chapter 8 the study of adsorbed layers on solid surfaces will be considered.

## 7.2 Surfaces of single crystals

A single crystal of a material can be cut in any direction, but certain cuts will expose a particular crystal plane. For most materials of interest, it is necessary to only consider a few crystal structures: these are simple cubic (sc), face centred cubic (fcc), body centred cubic (bcc), and hexagonal close packed (hcp). Directions within crystallographic lattices are described by Miller indices, in which a set of three (sc, bcc and fcc) or four (hcp) integral numbers are used. The way in which Miller indices are defined can be understood by considering Figure 7.3. The circles represent atoms in a sc lattice, and three Cartesian axes are defined. The unit cell has dimensions $(a, b, c)$ in the directions $x$, $y$, and $z$ respectively. In order to define the indices, the following steps are taken:

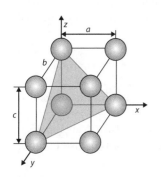

**Figure 7.3.** Planes defined in a simple cubic lattice. The (111) plane is shaded.

1. Determine, in multiples of the unit cell dimensions, where the plane cuts each axis. In the figure, the shaded plane intersects the axes at 1, 1, 1. Note that if a plane runs parallel to an axis, the intercept will be infinity. Also, it is possible to end up with negative numbers and fractions.

2. Take the reciprocal of the numbers determined in step 1. In the example given, the Miller indices are (111). Parentheses are traditionally used to denote a plane, while square brackets are used to describe directions (which are perpendicular to the corresponding planes). Clearly if one or more of the intercepts is infinity, the corresponding Miller index is zero. If one of the values is negative, it is written with a bar over the number, as in [$\bar{1}$10].

The low-index planes are the most important for our purposes, and three planes for the fcc structure are shown in Figure 7.4. Solid substrates, such as silicon, are grown and cut so that specific crystal faces are available for use in

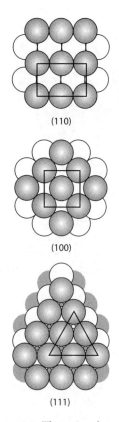

(110)

(100)

(111)

**Figure 7.4.** The major planes in the face centred cubic (fcc) system. The atomic layers are shown as light blue (top layer), white (second layer), and dark blue (third layer).

the semiconductor industry, and are also useful as substrates for film deposition and analysis. For the reasons discussed below, the choice of crystal face is important in many cases, as the energetics of the surface and therefore the way it interacts with adsorbates and deposited films depends on the structure of the face.

It is important to realize that the formation of a surface causes stress to a material. Atoms at the surface will not be bonded to as many neighbours as atoms in the bulk are – that is, the coordination is affected. For example, an atom on the [100] surface of an fcc crystal has only eight nearest neighbours, compared to 12 in the bulk. The loss of coordination, and therefore increase in surface energy, is compensated by changes in the structure in the surface region, and the greater the change in coordination, the greater the extent of change. This phenomenon is known as **surface relaxation**, and it is often seen as a change in spacing of the atomic layers within a certain distance of the solid–vacuum or sold–gas interface. The part of the surface that is affected is called the **selvedge**, and is typically 5–6 atomic layers thick.

In some circumstances, the surface undergoes a gross change in structure known as **surface reconstruction**, leading to a structure that is fundamentally different to that which would be expected if the bulk structure were suddenly terminated. This occurs when the surface energy is particularly high, and is manifested as a change in periodicity from what is expected. This is particularly likely to occur in semiconductors, where the electron density tends to be more localized than in metals, but reconstructions are also seen on metal surfaces, such as gold. An example of a surface reconstruction is the so-called $(2 \times 1)$ reconstruction of Si(100), shown in Figure 7.5.

Direct visualization of reconstructions is possible with the high resolution methods now available for studying surface structure, such as atomic force microscopy, described in Section 7.3.2. A classic image is that of the $(7 \times 7)$ reconstruction of Si(111), shown in Figure 7.6. Note that in this image it is possible to identify two defects.

It is important to realize that the adsorption process also affects the energetics of the surface, and adsorbate-induced surface relaxation also occurs.

## 7.3 Techniques for studying solid surfaces

In recent years a number of techniques that have dramatically improved our understanding of the physics and chemistry of surfaces have been developed and applied. These can be loosely grouped into the categories of spectroscopic and diffraction methods, involving interaction of matter with radiation or electrons, and scanning probe microscopy methods.

### 7.3.1 Spectroscopic and diffraction methods

In general, the spectroscopic methods give information about the *identity* (what atoms are there) and *chemistry* (how are they bonded together) of the

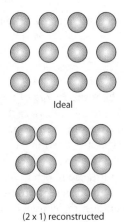

Ideal

(2 x 1) reconstructed

**Figure 7.5.** Top view of the $(2 \times 1)$ reconstruction of Si(100). The perturbation extends down into the surface to the fourth layer.

surface. Surface diffraction methods give information about the *ordering* and *structure* (how atoms are arranged) in the surface layers.

### X-ray photoelectron spectroscopy (XPS)

Also known as *electron spectroscopy for chemical analysis* (ESCA), this method is one of the most versatile and powerful of the spectroscopic techniques. It was developed in the 1950s by the Swedish scientist Kai Siegbahn (who won the Nobel Prize for his discovery in 1981) and commercialised in the following decade. A beam of X-rays impinging on a surface causes the ejection of electrons from core levels in the atoms due to the photoelectric effect (Figure 7.7), provided the energy of the X-rays is sufficient to overcome the energy holding the electrons in proximity to the nucleus (the binding energy). When a monochromatic source of X-rays of known energy is used, the binding energy can be determined if the kinetic energy of the electrons is measured.

**Figure 7.6.** The Si(111) $7 \times 7$ reconstruction. From Wiesendanger *et al.* (1990).

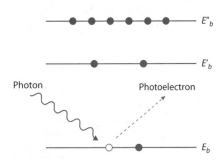

**Figure 7.7.** The process of X-ray photoelectron spectroscopy.

The kinetic energy of the ejected electrons is equal to the difference between the energy of the X-rays and the binding energy, $E_b$:

$$E_k = h\nu - E_b + \phi \qquad (7.1)$$

in which $\nu$ is the wavelength of the incident photons, $E_b$ is the energy of the core level from which the electron is ejected (Figure 7.8) and $\phi$ is known as the spectrometer work function, a characteristic of the instrument used to perform the measurement. As Figure 7.8. shows, the energies are determined by the quantum numbers $n$ ($1, 2, 3, \ldots$ which designates the shell: $K, L, M$, etc.), $l$ ($0, 1, \ldots, n-1$ designating the s, p, d, $\ldots$ orbitals) and the spin–orbit coupling constant $j$ ($j = l \pm 1/2$). Because the energy levels for different atoms, measured relative to the vacuum (reference) level are unique, the emission of electrons from different atoms will result in electrons detected at well-defined energies, and the resulting spectrum allows the determination of the composition of the surface. Peaks in an XPS spectrum are labelled according to the shell from which the electron is removed in the ionization process (see Table 7.1). A broad XPS scan, covering all the binding energy range, is known as a *survey spectrum* or *survey scan*, and is a useful way of identifying what elements are present at the surface of a given sample. An example survey scan

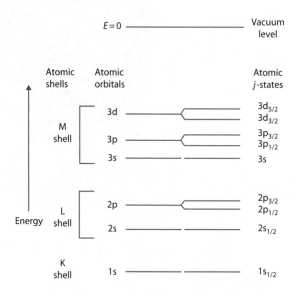

**Figure 7.8.** Energy level diagram of atoms in a solid.

**Table 7.1.** Notation for XPS and Auger spectra.

| XPS notation | $1s_{1/2}$ | $2s_{1/2}$ | $2p_{1/2}$ | $2p_{3/2}$ | $3s_{1/2}$ | $3p_{1/2}$ | $3p_{3/2}$ | $3d_{3/2}$ | $3d_{5/2}$ |
|---|---|---|---|---|---|---|---|---|---|
| Auger notation | $K$ | $L_1$ | $L_2$ | $L_3$ | $M_1$ | $M_2$ | $M_3$ | $M_4$ | $M_5$ |

of a fluorinated copolymer, fluorinated ethylene propylene (FEP), which contains only carbon and fluorine, is shown in Figure 7.9.

XPS is a highly surface-sensitive method, primarily due to the short mean free path of electrons in the solid. Although the penetration depth of X-rays of energy of, for example, 1 keV is approximately 1 μm, electrons of the same energy have a mean free path of only 2.5 nm (Figure 7.10), and therefore only electrons released from atoms in the first few layers will be released and detected. Electrons released from atoms deeper than this will either be absorbed before release, or will have a different energy and will contribute to the background noise.

Most modern XPS instruments are fitted with an ion gun that can bombard the surface with accelerated ions of argon and etch away the surface layers in a controlled manner. By successively etching and measuring the signal, information about the relative concentrations of elements as a function of

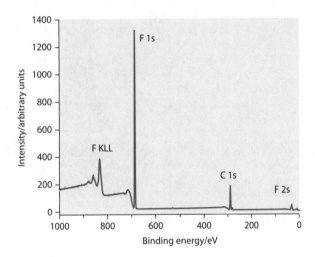

**Figure 7.9.** XPS survey spectrum of a fluorinated polymer. Peaks due to carbon and fluorine can be clearly seen. The peak labelled KLL arises from the Auger process, described later in this section. (Data of B. J. Wood.)

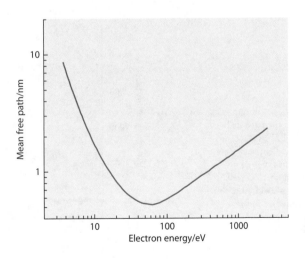

**Figure 7.10.** The inelastic mean free path of electrons in a solid. (Data from Somorjai, 1994.)

depth can be gained. This process is known as **depth profiling**. Figure 7.11. shows a beautiful example of depth profiling, where survey spectra were run after successive etching steps on a sample of glass coated with tungsten and tin oxides.

The other essential feature of XPS that contributes to its usefulness is the fact that, when measured at high resolution, the energies of the detected electrons are dependent not only on the identity of the atoms, but on their local chemical environment. For example, the binding energy of a carbon atom bound only to other carbon atoms is close to 285 eV whereas the 1s carbon signal from a C=O carbon is detected at approximately 289 eV. When bonded to a very electronegative atom like fluorine, the shift can be as large as 9 eV. A modern spectrometer with a monochromatic source can easily distinguish differences of these magnitudes (Figure 7.12), and therefore it is possible to gain information about the bonding of the

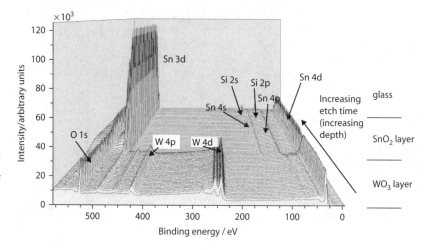

**Figure 7.11.** Survey scans of a coated glass sample as a function of etch time, showing the transition from a surface layer of $WO_3$, beneath which a layer of $SnO_2$ becomes apparent, followed by the bulk glass. (Courtesy of Kratos Analytical.)

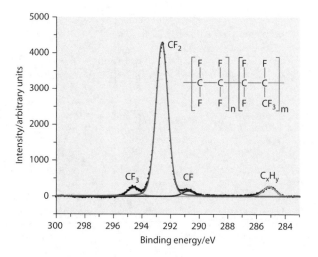

**Figure 7.12.** High-resolution XPS spectrum of the carbon 1s region from the FEP surface, as in Figure 7.9. It is clear that there are three different types of carbon in the polymer, as well as a fourth species associated with carbon contamination on the surface. Such 'adventitious carbon' is ubiquitous. (Data of B. J. Wood.)

surface atoms. The separation of the peaks is determined by a number of instrumental factors, but through curve fitting it is possible to detect peaks separated by about 0.5 eV. The variation in binding energy with chemical bonding is known as the chemical shift. It arises from small variations in the energies of the core levels due to charge shifts on bonding. Not surprisingly, a change in oxidation state also causes a large chemical shift.

Figure 7.13. shows a diagram of the layout of a modern XPS instrument. Note that like all electron spectroscopic techniques, the measurement must be performed under ultrahigh vacuum, to prevent scattering and absorption of the photoelectrons by air.

The relative intensities of peaks in a spectrum are related to the relative concentrations of atoms in the surface region. However, the probability of emission occurring from a given core level of a given atom differs according to a parameter known as the photoemission cross-section, which has been tabulated. In practice, it is normal to measure relative sensitivity factors for a given instrument, using standards that are as closely matched to the sample as possible, and these can be used in conjunction with the integrated areas under peaks to determine relative concentrations of different species in the surface.

### Auger electron spectroscopy (AES)

Based on a process identified by Pierre Auger in 1923, Auger spectroscopy involves the detection of electrons produced by the process illustrated in Figure 7.14. A photon, or more commonly an electron, incident on the surface causes the ejection of a primary photoelectron. The excited ion produced by this ionization process can relax by one of two mechanisms: ejection of a photon (which is known as X-ray fluorescence) or the Auger process, as shown in Figure 7.14. In the Auger process, the hole formed by the ejection of the core electron is filled by the relaxation of another electron in a

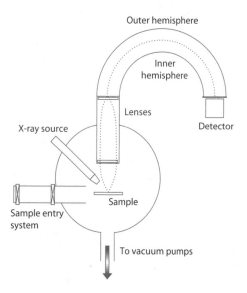

**Figure 7.13.** Layout of a modern XPS spectrometer. The instrument depicted here uses a hemispherical sector analyser.

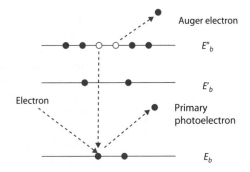

**Figure 7.14.** The Auger process.

### The Meitner effect?

The history of the Auger effect has interesting ties to the Second World War and nuclear fission. Although named after French scientist Pierre Auger, the effect was first discovered by the Austrian-born Jewish physicist Lise Meitner in 1923. Meitner worked closely with chemist Otto Hahn for 30 years, and together they studied radioactivity in the Kaiser Wilhelm Institute for Chemistry in Berlin. In 1918 they discovered the element protactinium. In 1938 Meitner was forced to flee Germany to Sweden, where she continued her work. She met clandestinely with Hahn in Copenhagen in November of that year to plan experiments, performed in Hahn's laboratory and published in January 1939 that provided evidence for nuclear fission. Meitner published the physical explanation for the process and named it nuclear fission in February 1939. However, in 1945 Hahn was awarded the Nobel prize while Meitner was overlooked. As some compensation, in 1966 Hahn and Meitner were awarded the Fermi prize, and element 109 is named meitnerium in her honour.[6] The Hahn Meitner Institute in Berlin is today an important research facility.

radiationless transition, leaving the atom in an excited state. The energy of the excited state is relieved by the ejection of a further electron, the **Auger electron**, which is the one that is detected. The other two electrons (incident and primary) contribute to the background.

A distinctive feature of Auger spectroscopy is that the energy of the Auger electron is independent of the energy of the exciting electron or photon. This is useful for identifying Auger peaks that are often seen in XPS spectra – if the spectrum is remeasured with a different excitation energy, the Auger peaks will remain at the same apparent binding energy whereas the XPS peaks will move, being dependent on the energy of the incident photons.

Because the ionizing source is usually an electron beam, and three electrons are involved in the process, the background is quite large compared to the peaks. For this reason, it is common to plot the derivative of the number of electrons detected $(dN(E)/dE)$ as a function of binding energy, which

6. source http://www.malaspina.com/site/person_623.asp

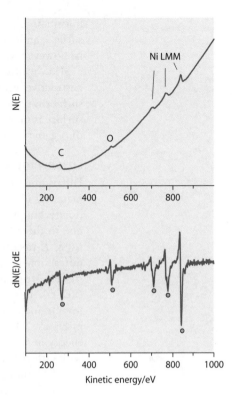

**Figure 7.15.** Auger spectrum of Ni metal, plotted as raw intensity, $N(E)$ (upper curve), and as a derivative spectrum, $dN(E)/dE$ (lower curve). The suppression of the background can be clearly seen in the lower spectrum. Filled circles indicate the points at which the peak positions are measured. (Data of B. J. Wood.)

emphasizes the peaks and causes the background to largely disappear. By convention, the peak position is defined as the minimum of the main Auger feature. Plotting the data in this manner has the effect of emphasizing small peaks measured on a large sloping background (Figure 7.15).

Using an electron beam as the source of energy has another benefit, in that the beam may be focussed to a very small spot (of the order of 10 nm diameter). If the spot is scanned over the surface, it is possible to study the distribution of elements on a surface to high spatial resolution. An instrument specifically designed to exploit this property is known as a scanning Auger microprobe.

Auger transitions are denoted by three capital letters, such as KLL, where the first letter refers to the shell (principal quantum number) from which the initial ionization took place, the second refers to the shell from which the electron dropped to replace the ionized electron, and the third refers to the shell from which the Auger electron is ejected. Subscripted numerals are used to indicate subshells, such as $KL_2L_3$ (see Table 7.1).

A significant disadvantage with the Auger technique is the fact that the electron beam tends to degrade polymeric and organic materials.

### Secondary ion mass spectrometry (SIMS)

When the surface of a solid is bombarded with ions at high energy, atoms and molecular fragments are gradually removed from the upper layers, and under the right conditions the resulting ions can be analysed for their masses. The information gained by this method is a measure of the composition of the surface, and it is particularly sensitive to very small levels of impurities. SIMS is, however, very poorly quantifiable.

SIMS measurements are made in one of two ways. *Dynamic SIMS* is a destructive technique where the bombardment rate is relatively high and new surface is constantly being exposed. In *static SIMS* the rate of damage to the surface is much slower. This method is very good for identifying trace levels of organic contaminants and studying organic-modified surfaces.

### Low energy electron diffraction (LEED)

Diffraction techniques are useful ways of determining the structures of materials in which the atoms are arranged in an ordered way. Most diffraction techniques, such as X-ray diffraction, are not inherently surface sensitive due to the relatively high penetration ability of X-rays. There are ways of using X-rays for surface sensitive diffraction (see grazing incidence X-ray diffraction in the next section), but the more commonly used method for surface structure determination is that of low energy electron diffraction (LEED). In this method electrons are back scattered elastically from the sample and analysed as a function of angle in apparatus shown schematically in Figure 7.16. The surface sensitivity arises from the fact that electrons of energy in the range 20–1000 eV have a very short mean free path in solids, with little dependence on the actual identity of the material. According to the so-called 'universal curve' of the inelastic mean free path of electrons in solids (Figure 7.10), electrons in this energy range travel approximately 0.5–1.5 nm into the solid, so the scattered electrons reveal information about the structure of only the top few atomic layers.

In the LEED method, a monochromatic beam of electrons impinges on the sample. Those electrons that are backscattered from an ordered array of atoms in the sample emerge as discrete beams at certain angles, and are detected by a phosphor screen after passing through a number of grids at various potentials that ensure only elastically scattered electrons are detected. A description of the theory of the diffraction of electrons from surface layers is beyond the scope of this book, and excellent descriptions are available. The

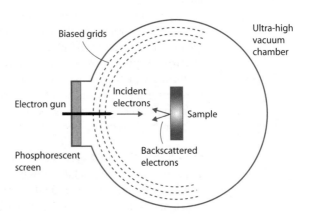

**Figure 7.16.** LEED apparatus (modified from Kolasinski, 2002).

analysis of diffraction patterns involves interpreting the positions and intensities of the diffracted beams, and provides the following information:

- The crystal structure of the surface layers, including unit cell size and symmetry, is obtained from the positions of the diffracted beams.
- The coordinates of the atoms within the unit cell, deduced from the intensities of the beams and their dependence on the energy of the incident beam.

Like any diffraction technique, the process of determining the geometric structure of the surface from the LEED spectrum is complex. In fact, because of complications such as multiple scattering, where electrons scatter elastically from more than one atom before being detected, it is not feasible to move directly and unambiguously from the LEED spectrum to the structure. The way this is resolved in practice is to postulate a trial structure, from which the predicted LEED spectrum can be calculated and compared with the measured spectrum. In an iterative process, the postulated structure is then refined, until the 'best fit' with experimental data is obtained.

### Grazing incidence X-ray diffraction

The technique of grazing incidence X-ray diffraction (GIXD) has been discussed earlier in the context of structural studies of organic thin films, such as LB films and self-assembled monolayers, on substrates (see Section 5.8). It is also possible to take advantage of the surface sensitivity of GIXD for the study of other types of solid surfaces where the surface structure may differ from the bulk. For example, ion implantation is one form of surface modification where GIXD can provide useful information about structural change. The method has also been widely used to examine residual stress in surface layers and the effect of various surface-altering treatments such as wear and sputtering.

## 7.3.2 Scanning probe techniques

The first of the scanning probe methods, scanning tunnelling microscopy (STM) was discovered in 1982 by Gerd Binnig and Heinrich Röhrer working at the IBM Zürich Research Laboratory, and earned them the Nobel Prize just four years later. This development spurred the growth of a large family of methods for understanding surface topography, composition, structure and mechanical properties. All scanning probe microscopy (SPM) methods are based on a probe being raster scanned over a surface, and the interaction between probe and surface being measured. The various methods differ in the way that the probe and sample interact. Of the many variations of SPM, arguably the most useful are atomic force microscopy (AFM) and STM. These are described briefly below:

### Atomic force microscopy (AFM)

By far the most commonly used scanning probe method, AFM relies on a sharp tip mounted on a flexible cantilever as a probe. In the most common

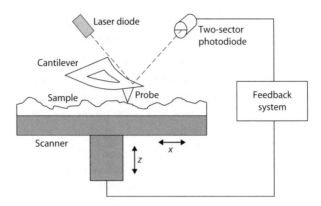

**Figure 7.17.** Basic experimental arrangement for atomic force microscopy.

mode of operation, called contact mode, the tip is brought into light contact with the surface, and scanned in a raster pattern. If it is scanned at a constant height (that is, $z$-value; see Figure 7.17), the deflection of the cantilever changes, depending on the height of the surface under the tip at any given moment. This deflection can be measured by a sensitive optical sensor, consisting of a laser diode which reflects from the back of the cantilever and the reflected beam position sensed by a two-sector photodiode. In practice, the measurement is often done the other way around: the sample is mounted on a piezoelectric scanner, which can move it in three dimensions extremely precisely, and the tip is fixed while the sample is scanned. In the $z$-direction, a feedback mechanism is used in conjunction with the output from the optical sensor to maintain a constant deflection of the cantilever as the sample is scanned. In this mode, called constant force mode, the measured movement of the sample required to maintain constant force is mapped, to give a representation of the topography (that is, the variations in height over the surface of the sample). Under favourable circumstances, images can be obtained at atomic resolution, even at room temperature in air or under liquids.

The result of an SPM scan is a two-dimensional topographic map of the surface, such as that shown in Figure 7.6. It should be emphasised that the interpretation of AFM images needs to be done carefully – in reality, the image is not purely a function of the sample surface, but represents a convolution of the forces between the tip and the surface. It also represents a very small section of a macroscopic surface, which is not necessarily representative of the entire sample. Despite these facts, AFM has revolutionized the study of surfaces at high resolution, and is an incredibly useful and versatile tool. Its popularity can be attributed to a number of factors: the ability tó operate in air and even under liquids; the ability to use a number of different types of probes to gain different information and the very large number of variations on the basic methods that have been developed are some reasons. Some modes of operation are listed below:

### Non-contact mode (NCM)

Rather than being in physical contact with the surface, the tip is held a small distance ($\sim 5$ nm) above the surface, and is oscillated using a piezoelectric modulator on the cantilever. The characteristic frequency of oscillation of the cantilever is modified by the proximity of the surface, and this information can be used to keep the distance between the tip and surface constant by a feedback mechanism, and therefore give a measure of the surface topography. The major advantage is that soft surfaces can be imaged with less risk of affecting the surface during the measurement.

### Lateral force microscopy (LFM)

As the tip is scanned, a four-sector photodiode is used to detect not only forces on the tip in the vertical ($z$) direction as in normal AFM, but also measures the lateral (sideways) forces which arise due to the friction felt by the tip as it moves relative to the surface. LFM is able to give more information about the chemical nature of the surface, as areas which are different chemically may offer different lateral resistance to the tip. This mode also tends to enhance contrast of steps in the direction perpendicular to the scan direction.

### Magnetic force microscopy (MFM)

If the tip is coated with a ferromagnetic material, there will be a magnetic interaction between tip and surface simultaneously with the normal atomic force interaction that is always present. This is commonly used to image magnetic domains.

### Scanning tunnelling microscopy (STM)

In STM, the tip/cantilever mechanism is replaced with a sharpened metal tip connected to a source of bias voltage (typically $\pm 2\,V$ with respect to the sample surface). When the tip is brought to within 2–5 Å of a conducting surface, electrons tunnel across the gap. The current is exponentially dependent on the distance between the two (typically changing by a factor of 10 for a 1 Å change in separation), and this extreme dependence is the key to the resolving power of STM – if one atom on the tip protrudes beyond all the others, the majority of the current will tunnel to this atom, allowing true atomic resolution imaging.

Although STM images are often interpreted as topographic representations of the surface, it can be shown that under conditions of small bias and constant current, an STM image corresponds to a map of constant local density of states at the Fermi level. Interpretation must therefore be done very carefully, in that heterogeneous surfaces may appear to have topographic features which are in fact a result not of physical height variations, but of variations in electronic structure. However, this feature can also be beneficial, in that STM can give useful information about the surface electronic structure.

Although it is necessary for the sample to be conducting to perform STM imaging, it is possible to obtain good images of thin non-conducting samples such as thin films on, for example, graphite substrates. This has extended the usefulness of STM as an important method, which nicely complements the other SPM techniques.

## SUMMARY

Solid surfaces are very different to the bulk of the same material. As well as the impurities and defects that are nearly always present, the atoms of the surface are energetically different to the bulk atoms, and this affects their properties and reactivity. Directions within crystals are described by Miller indices, and a crystal can be cut so that one face of the crystal makes up the surface of interest. The formation of a surface causes stress due to the changes undergone by the atoms in the immediate region, known as the selvedge, leading to such phenomena as surface relaxation and surface reconstruction.

A number of sensitive methods for studying solid surface structure and composition have been developed. Some, such as X-ray photoelectron spectroscopy, Auger spectroscopy and secondary ion mass spectrometry, give primarily information about the composition and bonding of the surface. Diffraction methods such as low energy electron diffraction and grazing incidence X-ray diffraction yield information about the structure – the arrangement of the atoms in the top few molecular layers. The complementary methods classed as scanning probe microscopy techniques are important methods of determining a range of information about the outermost part of the surface with up to atomic spatial resolution.

## FURTHER READING

Attard, G. and Barnes, C. (1998). *Surfaces*. Oxford University Press, Oxford.

Briggs, D. and Seah, M. P. (1990). *Practical Surface Analysis, 2nd edn*. John Wiley, Chichester.

Kolasinski, K. W. (2002). *Surface Science: Foundations of Catalysis and Nanoscience*. John Wiley, Chichester.

Moulder, J. F., Stickle, W. F., Stobol, P. E., and Bomben, K. D. (1992). *Handbook of X-ray Photoelectron Spectroscopy*. Perkin Elmer Corporation, Eden Prairie.

Somorjai, G. A. (1994). *Introduction to Surface Chemistry and Catalysis*. John Wiley, New York.

## REFERENCES

Wiesendanger, R., Tarrach, G., Scandella, L., and Güntherodt, H.-J. (1990). *Ultramicroscopy*, **32**, 291.

## EXERCISES

**7.1.** For each of the simple cubic structures shown below, determine the Miller indices of the plane indicated.

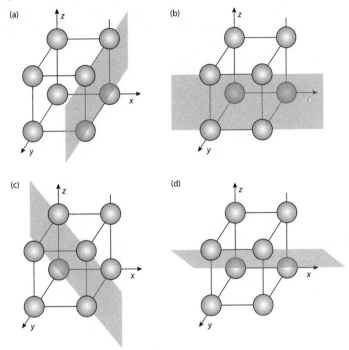

(a)

(b)

(c)

(d)

**7.2.** Consider the XPS spectrum shown below. Adventitious carbon would be expected to give a 1s peak at 284.8 eV. Identify the origin of the individual peaks and the identity of the metal which gave rise to the spectrum. You will need to use tables in the *Handbook of X-ray Photoelectron Spectroscopy* or similar reference work (see Further Reading list).

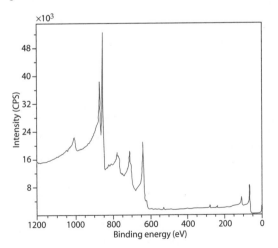

**7.3.** How many carbon 1s peaks would you expect to observe (assuming they could be resolved) in the XPS spectrum of each of the following linear polymers?

$-(-CF_2 -CF_2-)_n-$
poly(tetrafluoroethylene)

$-(-CH_2 -CH_2-)_n-$
polyethylene

$-(-CF_2 -CH_2-)_n-$
poly(vinylidene fluoride)

$-(-CH_2 -CH_2-O-)_n-$
poly(ethylene glycol)

$-(-CH_2 -CH_2-CH_2-O-)_n-$
poly(propylene glycol)

# 8 Adsorption at the gas–solid interface

## 8.1 Introduction

A solid surface in contact with a gas usually attracts an adsorbed layer of gas molecules. Even a gas like nitrogen, which is non-polar and not very reactive, will adsorb to a surface under certain conditions. When adsorption occurs the process is spontaneous which means that the free energy change, $\Delta G$, is negative. In most cases there is also a decrease in entropy, $\Delta S$, because the gas loses degrees of freedom. Hence, to ensure that $\Delta G$ is negative, the enthalpy change, $\Delta H$, must be negative and large enough to outweigh the entropy term:

$$\Delta G = \Delta H - T\Delta S.$$

The adsorption of gases plays an important role in many processes, in particular heterogeneous catalysis. It is the weakening or breaking of chemical bonds that occurs upon adsorption that makes some solid surfaces effective catalysts for gas phase reactions. Another important use of gas adsorption is for the characterization of porous materials, where information about the surface area and nature of the surface can be gained from measurements of the adsorption as a function of temperature and pressure. In this chapter we discuss the phenomenon, and see how it can be used as a means of characterizing materials and to help understand the important process of heterogeneous catalysis.

## 8.2 Physical and chemical adsorption

The adsorption of a gas (often called the adsorbate) on a solid surface (often called the adsorbent) may involve only physical interactions due to van der Waals forces, or there may be a chemical interaction with the formation of chemical bonds between the solid and the gas. These two processes are referred to as physical adsorption and chemical adsorption or chemisorption, respectively. The distinction between them is very important as the adsorption processes and the behaviour of the adsorbed layers are quite different.

Physical adsorption, because it involves only the forces of molecular interaction, is very similar to the condensation of a vapour to form a liquid. The enthalpy of physical adsorption is roughly the same as the enthalpy of liquefaction and in many ways the adsorbed material, especially when many layers have been adsorbed, behaves like a two-dimensional liquid.

In chemisorption, on the other hand, a chemical bond is formed. Consequently the enthalpies of adsorption are much greater than for physical adsorption, being comparable to those of bond formation. Furthermore, the adsorbed atoms are localized at particular sites on the solid surface and only one layer of adsorbate may be chemisorbed. Physical adsorption on top of a chemisorbed layer is possible if the conditions are appropriate.

### 8.2.1 Contrasts between physical and chemical adsorption

In comparing the characteristics of physical and chemical adsorption it is useful to use the analogies with condensation and chemical reaction. It will be seen that no single characteristic can be used to distinguish these two types of adsorption as there are often exceptions, so several characteristics should be examined.

#### Enthalpy of adsorption

The enthalpy of physical adsorption is usually no greater than a few times that of condensation, with values well below $20 \, \text{kJ} \, \text{mol}^{-1}$, whereas the enthalpy of chemisorption is usually much greater and is rarely less than $80 \, \text{kJ} \, \text{mol}^{-1}$. However there are occasional exceptions (e.g. for the chemisorption of hydrogen on glass the enthalpy is only about $12 \, \text{kJ} \, \text{mol}^{-1}$).

#### Specificity

As chemisorption involves a chemical reaction between gas molecules and groups on the surface of the solid it is highly specific. Physical adsorption, on the other hand, is non-specific and occurs with any gas–solid combination under the appropriate conditions of temperature and pressure.

### Reversibility

Physical adsorption can be reversed, i.e. the gas can be desorbed, by simply reducing the gas pressure and usually without raising the temperature. However, the process may be slow because of diffusion effects in, for example, powder samples. The desorption of a chemisorbed gas is usually very difficult as it requires the breaking of chemical bonds. Low pressure and high temperature are usually needed. An extreme case is the desorption of oxygen from charcoal, where the oxygen is removed as carbon monoxide and dioxide.

### Thickness of the adsorbed layer

As pointed out earlier, chemisorption is limited to one adsorbed layer but in physical adsorption there are usually several layers adsorbed. However physical adsorption may occur on top of a chemisorbed layer and this may obscure the nature of the adsorption of the first layer.

### Conditions of adsorption

The experimental conditions favouring physical adsorption and chemisorption are similar to those favouring condensation and chemical reaction respectively: for physical adsorption, low temperatures and pressures close to those for liquefaction; for chemisorption, high temperatures and a wide range of pressures.

## 8.3 Measurement of gas adsorption

Gas adsorption is usually measured by one of two broad methods: observing the decrease in the amount of gaseous adsorbate present after exposure of the gas to the solid (by volumetric or flow methods), or the use of a sensitive microbalance to measure the increase in weight of the solid after exposure to the gas.

### 8.3.1 Volumetric methods

Traditionally, gas adsorption has been measured by methods in which the change in volume of gas during adsorption is measured directly.

In principle, the apparatus is very simple, as Figure 8.1 shows.

The mercury reservoirs beneath the manometer and the burette are used to control the levels of mercury in the apparatus above. Calibration involves measuring the volumes of the gas lines and of the void space in the sample bulb. All pressure measurements are made with the right arm of the manometer set at a fixed zero point so that the volume of the gas lines does not change when the pressure changes.

The apparatus, including the sample bulb, is evacuated and the sample may be heated to remove any previously adsorbed gas. Helium is usually used for

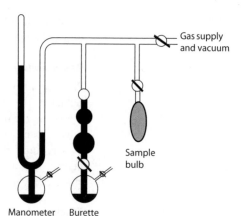

**Figure 8.1.** Basic volumetric apparatus for measuring the adsorption of a gas on a solid.

calibration as it is not appreciably adsorbed on solid surfaces under the usual conditions of measurement. With a set amount of helium in the apparatus, the volume is changed by a known amount using the calibrated burette and the consequent pressure change measured.

Nitrogen is normally used as the adsorbate when the surface area of a solid is to be determined. The sample bulb is immersed in liquid nitrogen for the calibration and for the measurement. With a fixed amount of nitrogen gas in the apparatus the drop in pressure when the tap to the sample bulb is opened is determined. Part of the pressure drop is attributable to the increased volume and the remainder to adsorption.

Measurements are made over a range of gas pressures and analysed by the linear form of the BET Eq. (8.6). The surface area is calculated from the monolayer capacity taking the area of a nitrogen molecule as $0.162\,nm^2$.

More modern variants of the basic apparatus shown above can be easily constructed, with improvements in vacuum taps and pressure sensing devices. It is no longer necessary to use mercury to alter the volume of the system, but a series of calibrated flasks with individual taps can be used to perform the same function. Pressure measurement is conveniently performed with an electronic pressure transducer, such as those which measure the variation in capacitance between two plates, one of which is a flexible diaphragm in contact with the gas.

### 8.3.2 Flow methods

Routine, automated measurement of gas adsorption is usually performed with the help of a flow system, where the adsorbate gas is mixed with a diluent and changes in concentration of the gas mixture occur as the gas is adsorbed on the surface (Figure 8.2). An example commonly used is a mixture of nitrogen in helium. When the sample is cooled to 77 K, the boiling temperature of nitrogen, adsorption occurs and the gas mixture flowing over the sample is depleted in nitrogen compared to that not reaching the sample. Such small changes in gas concentration can be measured using matched thermal conductivity detectors, located before and after the sample position. After

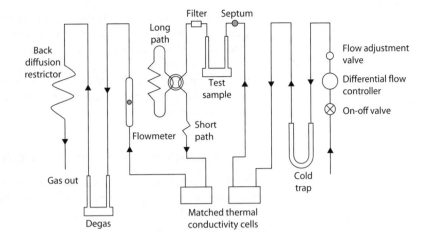

**Figure 8.2.** An example of a modern instrument for measuring gas adsorption, the Micromeritics Flowsorb system (courtesy of Micromeritics Corporation).

detector calibration, performed by injecting a known volume of nitrogen through a septum, the number of moles adsorbed can be determined. For a quick measurement of the surface area of a solid, it is often sufficient to measure a single gas concentration, but if a more detailed analysis of the surface is required, a multipoint analysis is performed using a number of different concentrations of the adsorbate gas in the mixture. The data are then fitted to one of the isotherm equations described in this chapter.

### 8.3.3 Gravimetric methods

A microbalance is just a very sensitive weighing device, usually comprising a modern form of the classic beam balance. Because only the difference between a counterweight and the sample is measured, microbalances can be sensitive to weight differences of nanograms. With such extreme sensitivity it is possible to measure the weight change caused by the adsorption of a single monolayer on a solid if the surface area is large. The normal procedure is to expose the sample to the adsorbate gas at a certain pressure, allowing sufficient time for equilibrium to be reached and then determining the mass change. This is repeated for a number of different pressures, and the number of moles adsorbed as a function of pressure plotted to give an adsorption isotherm (see Section 8.4).

Microbalances can be made to handle pressures as high as 120 MPa, so gases that adsorb weakly or boil at very low pressures can still be used. Such balances are usually made of stainless steel.

## 8.4 Adsorption isotherms

### 8.4.1 Introduction

The results of an adsorption measurement are usually plotted as an adsorption isotherm: the amount of gas adsorbed as a function of the equilibrium gas pressure. Often the pressure is referenced to the saturation pressure, $p°$, at the selected temperature. The saturation pressure is the gas pressure at which condensation would commence: in other words, the equilibrium vapour pressure of the liquefied gas at the selected temperature. It is common to use the relative pressure, which is the ratio of the actual gas pressure to the saturation pressure ($p/p°$), as the independent variable in adsorption isotherms.

### 8.4.2 Classification of adsorption isotherms

Understanding of adsorption at the gas–solid interface was greatly enhanced when Brunauer, Deming, Deming, and Teller (1940) realized that the numerous isotherms that had been reported to that time could be classified by their shapes into five groups. These shapes are shown in Figure 8.3, and the groups are termed Type I, Type II, ..., Type V.

The Type I isotherms are found when the gas is chemisorbed and limited to one layer of coverage. The other four types are considered to derive from physical adsorption. Types II and IV show steep rises in adsorption at low gas pressures suggesting a high affinity between gas and solid, whereas Types III and V exhibit low adsorptions at low gas pressures indicating relatively weak affinity between gas and solid.

At very low gas pressures all of the isotherm types approximate to a straight line. As this behaviour is reminiscent of Henry's law (solubility of a gas is proportional to its pressure) this section of the isotherms is often called the Henry's law region.

## 8.5 Isotherm equations

There have been a number of attempts to derive an equation that will fit the experimental adsorption isotherms and several of these have been reasonably successful. However the ability of an equation to fit the shape of an experimental isotherm does not provide a sensitive test for the model on which the equation was based. A more stringent test is to compare the enthalpy of adsorption deduced from the isotherms using the equation with the value measured calorimetrically, or to try to predict the isotherm for one temperature from the isotherm measured at another.

The following treatment will be limited to two of the most frequently used isotherms: that derived by Langmuir in 1918 for single layer adsorption and the extension of that equation to multilayer adsorption by Brunauer, Emmett, and Teller in 1938, known universally as the BET equation. Each of these equations can be derived in several ways, but only the simple kinetic derivations will be presented here. Only brief mentions will be made of other isotherm equations.

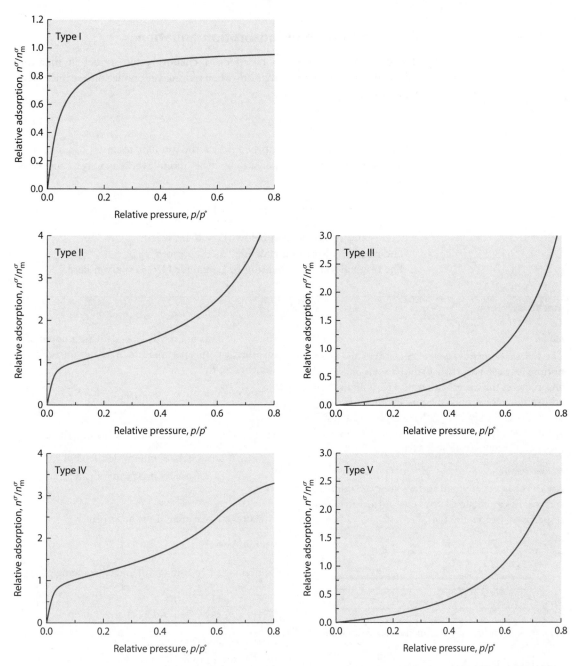

**Figure 8.3.** The classification of adsorption isotherms based on their shapes, proposed by Brunauer, Deming, Deming, and Teller. Isotherms on the left arise from strong interactions between gas and solid, those on the right from weak interactions. Adsorption values are shown as the ratio of the number of moles adsorbed, $n^{\sigma}$, to the amount required to cover the surface with a single monolayer, $n_{\mathrm{m}}^{\sigma}$.

### 8.5.1 The Langmuir adsorption equation

The Langmuir equation is based on a chemisorption model in that the adsorption is limited to a single adsorbed monolayer on the solid surface. The equation is:

$$\frac{n^{\sigma}}{n^{\sigma}_{\mathrm{m}}} = \frac{\alpha p}{1 + \alpha p} \tag{8.1}$$

where $n^{\sigma}$ is the amount adsorbed, $n^{\sigma}_{\mathrm{m}}$ is the amount required for complete monolayer coverage of the surface (the monolayer capacity), and $\alpha$ is defined by

$$\alpha = \frac{a_1}{b_1 \exp\left(-E_1/RT\right)} = \left(\frac{a_1}{b_1}\right) \exp\left(\frac{E_1}{RT}\right) \tag{8.2}$$

where $a_1$ and $b_1$ are defined below (Eqs. 8.1a, 8.1b).

The relative adsorption is defined as the ratio $n^{\sigma}/n^{\sigma}_{\mathrm{m}}$.

The original kinetic derivation of Langmuir (1915) is given here.

---

**Kinetic derivation**

*Model*

The kinetic derivation postulates a dynamic equilibrium between adsorbate in the gaseous phase and adsorbate in the absorbed layer. The limitation of adsorption to one monolayer means that adsorbing molecules can only attach themselves to bare surface, as adsorption on to an occupied site would produce a second adsorbed layer.

*Nomenclature*

Areas: $A$, $A_0$, $A_1$ are total surface area, area of bare surface, area occupied by one adsorbed layer respectively (see Figure 8.4).

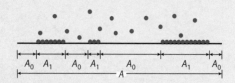

**Figure 8.4.** Nomenclature used in kinetic derivation of Langmuir adsorption equation.

Enthalpy of desorption: $E_1$.

Adsorption factor: $a_1$ (see Eq. 8.1a).

Desorption factor: $b_1$ (see Eq. 8.1b).

*Derivation*

The rate of absorption will depend on the pressure of gas (adsorbate), $p$, and the area of solid (adsorbent) surface available for adsorption, $A_0$:

$$\text{rate of adsorption} = a_1 p A_0. \tag{8.1a}$$

The rate of desorption will depend on the amount of adsorbed gas (proportional to $A_1$), and an activation energy, $E_1$.

$$\text{rate of desorption} = b_1 A_1 \exp(-E_1/RT). \tag{8.1b}$$

At equilibrium, area $A_0$ remains constant, therefore

$$\text{rate of ads. on } A_0 = \text{rate of des. from } A_1$$
$$a_1 p A_0 = b_1 A_1 \exp(-E_1/RT). \tag{8.1c}$$

The total surface area of the adsorbent is:

$$A = A_0 + A_1 \tag{8.1d}$$

Combining (8.1c) and (8.1d) and rearranging gives:

$$\frac{A_1}{A} = \frac{a_1 p}{b_1 \exp\left(-E_1/RT\right) + a_1 p} = \frac{\alpha p}{1 + \alpha p} \tag{8.1e}$$

where $\alpha$ is defined by Eq. (8.2) above.

As $n^{\sigma}$ is the amount of gas adsorbed, and $n^{\sigma}_{\mathrm{m}}$ is the monolayer capacity, then:

$$\frac{n^{\sigma}}{n^{\sigma}_{\mathrm{m}}} = \frac{A_1}{A}. \tag{8.1f}$$

Combination of (8.1e) and (8.1f) gives the Langmuir isotherm:

$$\frac{n^{\sigma}}{n^{\sigma}_{\mathrm{m}}} = \frac{\alpha p}{1 + \alpha p}. \tag{8.1}$$

The equation can be rearranged into a linear form:

$$\frac{p}{n^\sigma} = \frac{1}{\alpha n_m^\sigma} + \frac{p}{n_m^\sigma}. \tag{8.3}$$

Thus by plotting experimental values of $p/n^\sigma$ against $p$ a straight line should be obtained from which the values of $\alpha$ and $n_m^\sigma$ can be calculated.

## Assumptions of the Langmuir equation

As there are several derivations possible only those assumptions that arise from the basic model need to be considered. They are:

- The enthalpy of desorption, $E_1$, is constant and independent of the amount of surface covered. This implies that the surface is quite uniform and that the adsorbed molecules do not interact with neighbouring adsorbed molecules. Neither of these propositions holds in practical situations, but in some cases they are reasonable approximations.

- The molecules are adsorbed at definite sites and are unable to move over the surface.

- Each site can only accommodate one adsorbate molecule (this is the limitation to one layer).

- The molecules are adsorbed without dissociation. However, dissociation is often a feature of chemisorption, the type of adsorption to which the Langmuir equation is most applicable.

## Applications of the Langmuir equation

Plots of the Langmuir equation are shown in Figure 8.5.

It is clear that the equation generates curves that fit the Type I shape and that the adsorption reaches a limit when one complete monolayer of gas has been adsorbed. Increases in the value of $\alpha$ and hence in the value of $E_1$ (Eq. 8.2) show increasing steepness in the isotherms at low pressures reflecting increasing affinity between adsorbate and adsorbent.

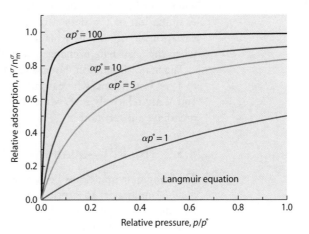

**Figure 8.5.** Plots of the Langmuir equation for different strengths of interaction between the gas adsorbate and the solid adsorbent surface.

Irving Langmuir, 1881–1957

Irving Langmuir was a happy man, with a harmonious family life, a love of the outdoors, and employment in the General Electric Research Laboratory that gave him the freedom to follow his intense interest in and curiosity about the natural world. He had a prodigious output of published research which together with unpublished wartime reports have been collected into twelve substantial volumes (Suits, 1961). The titles of these volumes indicate the breadth of his interests: 1. *Low Pressure Phenomena*; 2. *Heat Transfer – Incandescent Tungsten*; 3. *Thermionic Phenomena*; 4. *Electrical Discharge*; 5. *Plasma and Oscillations*; 6. *Structure of Matter*; 7. *Protein Structures*; 8. *Properties of Matter* (including *Interfacial Phenomena*); 9. *Surface Phenomena*; 10. *Atmospheric Phenomena*; 11. *Cloud Nucleation*; 12. *Langmuir – The Man and the Scientist* (a biography).

Langmuir's experimental methods were simple, though effective, and coupled with his exceptional analytical skill and insight, led to results which made major contributions to science and practical outcomes for his employer and others. In 1932 he received the Nobel Prize for Chemistry for his work in surface chemistry. There were several aspects to this award. One was based on the adsorption of gases on tungsten filaments which under the conditions of Langmuir's experiments was always chemisorption and led to the development of the Langmuir isotherm. Another was his work on insoluble monolayers where he developed a new method for measuring surface pressure and performed many experiments with what is now known as a Langmuir trough. It is entirely appropriate that the American Chemical Society should call its specialist journal on surface chemistry after him: *Langmuir*.

For data that fit the linear form of the equation (Eq. 8.3) values of the monolayer capacity can readily be obtained. If the size of the adsorbate molecules is known the value of $n_m^\sigma$ yields an estimate of the surface area of the solid.

Experimental data plotted according to the linear form of the Langmuir equation often give good straight lines. Frequently however, more detailed examination of the data reveals inconsistencies, such as excessively large variations in $n_m^\sigma$ and $E_1$ when the temperature is changed, or inconsistencies when the adsorbate is changed with the same adsorbent.

The major limitation, however, is that the Langmuir equation only applies to Type I isotherms, whereas most isotherms are of Types II and IV. Types III and V are relatively rare. Multilayer adsorption is much more common that monolayer adsorption.

### 8.5.2 The BET equation

The Langmuir model has been extended by Brunauer, Emmett and Teller to include multilayer adsorption, giving the equation now known as the BET equation. This equation has been used extensively in gas adsorption work. The original derivation was very similar to the kinetic derivation of the

Langmuir equation, but like the Langmuir equation, other derivations have also been developed. From these other derivations it has been concluded that the BET equation is valid for the model used. Any problems with the BET equation therefore arise from inadequacies in the model, not from the derivation. The BET equation is:

$$\frac{n^\sigma}{n^\sigma_m} = \frac{Zp}{(p^\circ - p)\{1 + (Z - 1)(p/p^\circ)\}} \tag{8.4}$$

where $Z$ is defined by

$$Z \approx \exp\{(E_1 - E_v)/RT\} \tag{8.5}$$

and is similar to $\alpha$ in the Langmuir equation, and the other terms are as defined earlier.

## Kinetic derivation

### Model

The kinetic derivation of the BET equation is an extension of the kinetic derivation of the Langmuir isotherm equation. It uses the Langmuir model without the limitation to monolayer adsorption. Additional assumptions are:

- There is no limit to the number of layers of gas that can be adsorbed (multilayer adsorption).

- The Langmuir equation applies to each layer.

- Adsorption and desorption can only occur at exposed surfaces.

- At equilibrium, the distribution of adsorbate between the different adsorption layers is constant, i.e. the areas covered by 0, 1, 2, 3, ... layers are constant.

### Nomenclature

The nomenclature is as used for the Langmuir equation, with extensions.

Areas: $A_0, A_1, A_2, A_3, A_4, \ldots, A_i$ refer to the areas covered by 0, 1, 2 layers, and so on.

Enthalpies of desorption: $E_1, E_2, E_3, E_4, \ldots, E_i$.

Adsorption factors: $a_1, a_2, a_3, a_4, \ldots, a_i$.

Desorption factors: $b_1, b_2, b_3, b_4, \ldots, b_i$

where the subscripts indicate the adsorption layer: $A_i$ is the area covered by $i$ layers, $E_i$ is the enthalpy of desorption from the $i$th layer, etc.

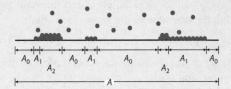

**Figure 8.6.** Nomenclature used in kinetic derivation of BET equation.

### Derivation

At equilibrium:

area $A_0$ remains constant, therefore

rate of ads. on $A_0$ = rate of des. from $A_1$

$$a_1 p A_0 = b_1 A_1 \exp(-E_1/RT) \tag{8.4a}$$

area $A_1$ remains constant, therefore

rate of ads. on $A_1$ + rate of des. from $A_1$
= rate of des. from $A_2$ + rate of ads. on $A_0$

$$a_2 p A_1 + b_1 A_1 \exp(-E_1/RT)$$
$$= b_2 A_2 \exp(-E_2/RT) + a_1 p A_0 \tag{8.4b}$$

and combination of (8.4a) and (8.4b) gives

$$a_2 p A_1 = b_2 A_2 \exp(-E_2/RT). \tag{8.4c}$$

In general, for the $(i-1)$th layer:

$$a_i p A_{i-1} = b_i A_i \exp(-E_i/RT) \tag{8.4d}$$

or

$$A_i = \{(a_i/b_i)p \exp(E_i/RT)\}A_{i-1}. \quad (8.4e)$$

The total surface area of the absorbent is

$$A = \sum_{i=0}^{\infty} A_i \quad (8.4f)$$

and the total amount of gas adsorbed is

$$n^\sigma = \frac{n_m^\sigma}{A} \sum_{i=0}^{\infty} iA_i \quad (8.4g)$$

From (8.4f) and (8.4g),

$$\frac{n^\sigma}{n_m^\sigma} = \frac{\sum_{i=1}^{\infty} iA_i}{\sum_{i=1}^{\infty} A_i}. \quad (8.4h)$$

Baly (1937) took the derivation to this point but could not evaluate the summations. Brunauer, Emmett, and Teller (1938) evaluated the sums by making the two simplifying assumptions that:

$$E_2 = E_3 = E_4 = E_5 \ldots = E_i = E_v \quad (8.4i)$$

where $E_v$ is the enthalpy of vaporization of the adsorbate, and

$$b_2/a_2 = b_3/a_3 = b_4/a_4 \ldots = b_i/a_i = C \quad (8.4j)$$

where $C$ is a constant.

Note that no assumptions have been made about the values of $E_1$ and $b_1/a_1$ for the first layer where there is direct interaction between adsorbate and the solid. The assumptions for the higher layers are equivalent to postulating that adsorption/desorption is the same as condensation/evaporation.

Define:

$$Y = (a_1/b_1)p \exp(E_1/RT) \quad (8.4k)$$

$$X = (1/C)p \exp(E_v/RT) \quad (8.4m)$$

$$Z = Y/X. \quad (8.4n)$$

These definitions, together with (8.4i) and (8.4j) simplify (8.4e) and its analogues:

$$A_1 = YA_0 = ZXA_0$$

$$A_2 = XA_1 = YXA_0 = ZX^2A_0$$

$$A_3 = XA_2 = YX^2A_0 = ZX^3A_0$$

and so on for other layers. For the general case ($i$th layer):

$$A_i = XA_{i-1} = YX^{i-1}A_0 = ZX^iA_0. \quad (8.4o)$$

Substituting (8.4o) into (8.4h) gives:

$$\frac{n^\sigma}{n_m^\sigma} = \frac{ZA_0 \sum_{i=1}^{\infty} iX^i}{A_0(1 + Z \sum_{i=1}^{\infty} X^i)}$$

$$= \frac{ZX}{(1-X)(1-X+ZX)}. \quad (8.4p, q)$$

The significance of $X$ is found by considering adsorption on a free surface at the saturation pressure, $p^\circ$. An infinite number of layers can then be adsorbed, so when $p = p^\circ$, $n^\sigma = \infty$, and to obtain this result from (8.4q),

$$X_{(p=p^\circ)} = 1.$$

From (8.4m):

$$X_{(p=p^\circ)} = 1 = (1/C)p^\circ \exp(E_v/RT).$$

Now

$$\frac{p}{p^\circ} = \frac{(1/C)p \exp(E_v/RT)}{(1/C)p^\circ \exp(E_v/RT)}$$

$$= \frac{(1/C)p \exp(E_v/RT)}{1}$$

$$= X. \quad (8.4r)$$

Thus, in (8.4p), $X$ can be replaced by the relative pressure, $p/p^\circ$, to give:

$$\frac{n^\sigma}{n_m^\sigma} = \frac{Zp}{(p^\circ - p)\{1 + (Z-1)(p/p^\circ)\}} \quad (8.4s)$$

which is the BET equation.

## Applications of the BET equation

The BET equation can be rearranged for linear plotting:

$$\frac{p}{n^\sigma(p^\circ - p)} = \frac{1}{Zn_m^\sigma} + \frac{Z-1}{Zn_m^\sigma}\frac{p}{p^\circ}. \quad (8.6)$$

Thus if the expression on the left is plotted against relative pressure $(p/p^\circ)$, a straight line should result if the data follow the BET isotherm. From such a straight line the values of $n_m^\sigma$ and $Z$ can be determined.

The value of $n_m^\sigma$ enables the surface area of the solid to be calculated if the area occupied by an adsorbate molecule is known.

From (8.4k)–(8.4n) it follows that:

$$Z \approx \exp\{(E_1 - E_v)/RT\} \tag{8.5}$$

which permits an evaluation of $E_1$.

Isotherm plots of the BET equation correspond well with the isotherm shapes found experimentally for Type II and Type III isotherms, as can be seen in the calculated graphs in Figure 8.7.

Low values of $Z$ give Type III isotherms and correspond to weak interaction between adsorbate and adsorbent and consequent low adsorption values at low relative pressures. As $E_1 \leq E_v$ adsorbing molecules will tend to adsorb on previously adsorbed molecules rather than on bare surface and it is therefore likely that the adsorbed film will consist of clusters of adsorbate several molecules thick with other areas left bare. Few isotherms of Type III have been reported so a general conclusion about the fit to the BET equation is not possible.

When $Z$ is large there is strong interaction between adsorbent and adsorbate and the BET equation gives isotherms of Type II. It is worth noting that the adsorption where the isotherms for high $Z$ start to flatten out is close to the monolayer capacity $(n^\sigma/n_m^\sigma = 1)$. Indeed, before the development of the BET equation it was often assumed that where the Type II isotherms began to flatten the adsorption corresponded roughly to monolayer coverage. The so-called 'point B' method for surface area determination was based on this approximation. ('Point B' was the name given to the point on the adsorption isotherm at the lower end of the nearly linear central section. Examination of the isotherm for $Z = 100$ in Figure 8.7 shows that the adsorption at this point is close to the monolayer capacity.)

Tests of Type II experimental data against the linear form of the BET Eq. (8.6) nearly always give a good linear plot in the intermediate range

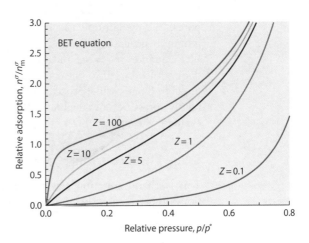

**Figure 8.7.** Calculated adsorption isotherms using the BET Eq. (8.4) with various values of $Z$.

of relative pressures: $0.05 < p/p° < 0.3$. At higher relative pressures there are usually deviations from linearity, but this can often be attributed to the presence of pores in the solid and consequent limitations on the number of layers than can be adsorbed. The $v$-layers isotherm, given below (Eq. 8.7), is an attempt to deal with this problem. At very low relative pressures any deviations from linearity are usually ascribed to inhomogeneity of the solid surface and consequent variations in the value of $E_1$ as the surface coverage in the first layer changes.

### Limitations of the BET theory

The BET theory suffers from many of the limitations found in the Langmuir theory. The absence of lateral interaction between adsorbed molecules suggests a somewhat unrealistic picture of the adsorbed film, sometimes described as the *battered metropolis* model: tall separate columns of adsorbed molecules. The BET treatment also assumes, unreasonably, that whenever a molecule is adsorbed, the full enthalpy of liquefaction is released irrespective of whether the molecule has neighbours in that position or not.

Attempts to improve the BET theory have generally meant the introduction of additional parameters which in most case cannot be evaluated independently. Thus any improvement in fitting experimental data can be attributed to the additional flexibility conferred by another parameter. Furthermore, most improvements destroy the simplicity of the BET equation and the ease with which the experimental data can be analysed to give values for $n_m^\sigma$ and $E_1$.

Calorimetric measurements of $E_1$ are not generally in agreement with values determined from the BET equation.

## 8.5.3 The BET equation for limited adsorption

If, instead of allowing adsorption to continue without limitation, a limit of $v$ layers is imposed the isotherm equation becomes:

$$\frac{n^\sigma}{n_m^\sigma} = \frac{ZX}{1-X} \frac{1-(v+1)X^v + vX^{v+1}}{1+(Z-1)X - ZX^{v+1}}. \tag{8.7}$$

Such a situation may occur for a porous material containing many very small pores, leading naturally to a limitation on adsorption capacity. This equation is of more general application than the usual BET equation as it reduces to the BET equation if $v = \infty$, and to the Langmuir equation if $v = 1$. However it does not lend itself to linear plotting so evaluation of the parameters is more difficult. One possible procedure is to take the adsorptions at moderately low relative pressures, use the linear BET Eq. (8.6) to evaluate $Z$ and $n_m^\sigma$, and then evaluate $v$ using Eq. (8.7).

## 8.5.4 The Freundlich and Temkin isotherms

This isotherm equation was proposed by Freundlich as an empirical relationship between adsorption and gas pressure:

$$n^\sigma = \beta p^{1/\kappa} \tag{8.8}$$

where $\beta$ and $\kappa$ are constants dependent on temperature and surface area. $\kappa$ is always greater than unity and is characteristic of the system being investigated.

The Freundlich equation gives adsorption isotherms that are similar to, but different from, those given by the Langmuir equation. In particular, the isotherm does not rise to a constant adsorption value as the gas pressure increases. The constants may be evaluated by expressing the equation in a logarithmic form and making the appropriate linear plot.

Although the Freundlich equation was developed as an empirical relationship, it has since been shown that it can be derived by modifying the assumption in the Langmuir model that $E_1$ is constant.

The Temkin isotherm may also be derived from the Langmuir model by allowing $E_1$ to vary with surface coverage. It may be written:

$$\frac{n^\sigma}{n_m^\sigma} = \frac{RT}{E_0\beta} \ln{(\kappa p)} \tag{8.9}$$

where $\beta$ and $\kappa$ are constants and $E_0$ is the enthalpy of desorption when $n^\sigma = 0$.

For chemisorption in particular, the Freundlich or the Temkin equation often gives a better fit to experimental data than does the Langmuir equation, due to the fact that enthalpies of adsorption usually fall with increasing adsorption.

### 8.5.5 Other isotherm equations

An alternative approach to the development of isotherm equations begins with the equation of state for a two-dimensional ideal gas. The assumptions made in deriving the equation of state drive the form of the isotherm equation. For example, if one begins with the ideal equation of state

$$\Pi\hat{A} = kT \tag{8.10}$$

one derives the isotherm

$$n^\sigma = Kp \tag{8.11}$$

where $K$ is a constant. This is the Henry's law (linear) behaviour common to all isotherms at low relative pressures. If more complex equations of state are used, correspondingly more complex isotherms result, which then incorporate the assumptions made in the initial equation of state. For example, the two-dimensional van der Waals equation contains empirical constants allowing for excluded area and intermolecular interactions, and therefore the resulting isotherm may be more appropriate at higher pressures than the corresponding ideal equation.

## 8.6 Capillary condensation

Condensation occurs when the actual vapour pressure exceeds the equilibrium vapour pressure. If the surface is curved, the Kelvin equation shows that this latter pressure may be significantly lower than $p^\circ$ ($\equiv p^\infty$). Thus condensation may occur at $p/p° < 1$, a phenomenon known as **capillary condensation** and not allowed for in the BET or $v$-layers treatments.

Combining Eqs. (2.11) and (2.8) gives the Kelvin equation in the form:

$$\ln\left(\frac{p^c}{p°}\right) = \left(\frac{\gamma \overline{V}^L}{RT}\right)\left(\frac{1}{r'} + \frac{1}{r''}\right). \tag{8.12}$$

Radii that lie in the gas phase are given a negative sign, so generally for an adsorbed film of liquid lining a capillary, both $r'$ and $r''$ will be negative and thus $p^c < p°$.

Capillary condensation can be expected to occur with porous solids when the relative pressure is high. It can also occur in the spaces between close-packed solid particles. In many cases the term $(1/r' + 1/r'')$ will differ for adsorption and desorption so the amount of gas on the surface at a particular pressure and temperature will depend on the history of the sample.

For example, the film of liquid in a capillary open at both ends (Figure 8.8) may take the form of a hollow cylinder when adsorption is increasing. Radius $r'$ would then be the capillary radius less the thickness of the adsorbed film, and $r''$, the radius in the plane containing the capillary axis, would be infinite.

Thus

$$\left(\frac{1}{r'} + \frac{1}{r''}\right) = \frac{1}{r'}. \tag{8.12a}$$

However, after the capillary has filled by capillary condensation and with the gas pressure now being reduced, desorption will take place from the roughly hemispherical menisci at each end of the capillary. Both radii will then be the same and approximately equal to the value of $r'$ in (8.12a). Thus

$$\left(\frac{1}{r'} + \frac{1}{r''}\right) = \frac{2}{r'}. \tag{8.12b}$$

In this and similar cases, capillary emptying will occur at a lower relative pressure than capillary filling and the isotherm will show adsorption–desorption hysteresis. Such isotherms are shown in Figure 8.9.

A notable example is the adsorption of water on silica gel.

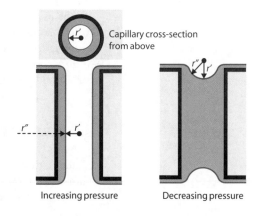

Capillary cross-section from above

Increasing pressure

Decreasing pressure

**Figure 8.8** Adsorption and desorption in an open-ended capillary: radii $r''$ are in the plane of the paper and radii $r'$ are in a plane perpendicular to the paper. Note that $r'' = \infty$ for the left diagram.

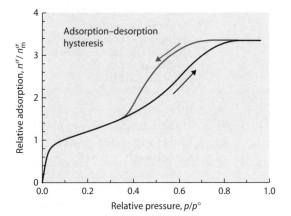

**Figure 8.9.** Adsorption–desorption hysteresis in the sorption of a gas on a porous solid.

## 8.7 Chemisorption

Studies of chemisorption are usually made under conditions chosen to minimize physical adsorption, such as higher temperatures and lower pressures than those used for physical adsorption work.

### 8.7.1 Adsorption isotherms

The conditions for chemisorption correspond to those used in the model for the Langmuir isotherm equation, so this equation could be expected to describe the experimental data. To some extent this is found to be the case, but there are complications.

The surface of even a pure solid is rarely smooth and in Chapter 7 we explored some of the defects that may be present. Such defects mean that the enthalpy of adsorption of a gas molecule does not have a fixed value but varies with the nature of the adsorption site. Gas molecules will be preferentially adsorbed on available sites with the largest enthalpies of desorption $(E_1)$ as these have the most negative values for the enthalpy of adsorption. Thus significant deviations from the Langmuir theory arise from the heterogeneous nature of nearly all solid surfaces: as adsorption proceeds the enthalpy of desorption will tend to decrease rather than remaining constant as required by the theory. There is, unfortunately, no universal pattern for this decrease and consequently no general theory can be developed. However, it is worth noting that the Freundlich and Temkin equations can be derived by modifying the Langmuir equation to accommodate particular forms of dependence of $E_1$ on surface coverage. Figure 8.10 illustrates, with a very simple model, some of the various interactions of a gas molecule with a solid surface and also indicates the further complication arising from interactions with previously adsorbed gas molecules.

By a special heat treatment it is possible to produce a charcoal with a nearly homogeneous surface ('Graphon[TM]').

**Figure 8.10.** Schematic diagram illustrating how surface heterogeneity can arise with a pure solid and showing different possible interactions between gas (o) and solid (●).

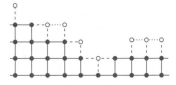

The Langmuir theory also assumes that the gas molecules are unchanged in the adsorption process, but in chemisorption the molecules are often dissociated.

In most cases the mechanism of chemisorption is more complex than the above treatment would imply. Usually when a gas molecule impacts on a solid surface and is not immediately reflected back into the gas phase it will be physically adsorbed. Even when the duration of this physical adsorption is brief, the molecule can move over significant distances and may find a chemisorption site to which it becomes attached. It should be noted that chemisorbed molecules are also able to diffuse on a surface without necessarily being desorbed, a fact which becomes important in the discussion of heterogeneous catalysis in the next section.

## 8.8 Heterogeneous catalysis

In heterogeneous catalysis the catalyst is in a different phase from the reactants, in contrast with homogeneous catalysis where the catalyst and the reactants are in the same phase. Typically, the catalysts for heterogeneous catalysis are solids, whereas a homogeneous catalyst may be a gas for a reaction between gas molecules or a solute for a reaction in solution. Thus heterogeneous catalysis usually involves the presence of a solid catalyst to increase the rate of a reaction in solution or in the gas phase. Here we will concentrate on gas phase reactions, but many of the principles can also be applied to reactions in the liquid phase.

Briefly, a catalyst is a substance that:

- increases the rate of a chemical reaction;
- does not itself take part in the overall chemical reaction;
- is unchanged chemically at the end of the reaction and the amount present is unaltered;
- is only required in small quantities to bring about a relatively large amount of reaction;
- does not affect the position of equilibrium in a reversible reaction.

The rate of a chemical reaction is governed by the height of the energy barrier between reactants and products, as shown in Figure 8.11.

The catalyst acts by lowering the height of the energy barrier which, according to the Arrhenius equation, increases the rate constant:

$$k = C \exp(-E^*/RT) \tag{8.13}$$

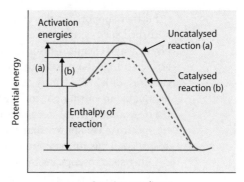

**Figure 8.11.** Schematic potential energy diagram for a chemical reaction showing the effect of a catalyst on the energy barrier: (a) represents the energy barrier for the uncatalysed reaction; (b) for the catalysed reaction.

Reaction coordinate

where $k$ is the rate constant, $C$ is a constant, $E^*$ is the activation energy (the height of the energy barrier).

## 8.8.1 Mechanisms of heterogeneous catalysis

In any gas reaction catalysed on a solid surface there are five distinct steps in the overall process. They are:

(a) diffusion of reactants to surface;

(b) adsorption of reactants;

(c) reaction within the adsorbed layer;

(d) desorption of products;

(e) diffusion of products away from surface.

Any of these processes may be the rate determining step in the overall reaction. However, by suitable experimental design the diffusion steps, (a) and (e), can be accelerated so they do not make any significant contribution to the overall kinetics. Interest thus centres on steps (b), (c), and (d), which will be discussed in turn.

### (b) Adsorption of reactants

The general principles underlying the adsorption of reactants have been discussed earlier in the treatments of physical and chemical adsorption (Section 8.2).

The first question to be considered concerns the nature of the adsorption: whether it must be physical or chemical for catalysis to occur. Faraday (1833) originally suggested that the main function of the catalyst is to provide, through physical adsorption alone, a higher concentration of the reactants and hence a higher reaction rate. There are two main arguments against this proposal. Firstly, it is now known that many catalysts are active at high temperatures where physical adsorption is negligible. Secondly, contact catalysts are often highly specific, whereas physical adsorption is a general phenomenon.

It follows, therefore, that heterogeneous catalysis always involves chemisorption, although physical adsorption may occur as well.

Further support for this conclusion has come from infrared spectra of the adsorbed species. In most cases this technique is able to distinguish between physical and chemical adsorption (provided that the chemisorbed species is IR active so that there is an observable change in the spectrum) and in many cases is able to identify the chemisorbed species. Such information is invaluable in sorting out the mechanism of the reaction.

For example, in the decomposition of formic acid on supported nickel, the IR spectra have shown that the formate ion is produced when the acid is adsorbed. Thus formate ion is the reaction intermediate.

Other spectroscopic techniques, such as UV–visible and NMR, may also be used in such investigations. In addition, there is now an array of other techniques available for studying surfaces, such as low-energy electron diffraction (LEED), secondary ion mass spectrometry (SIMS), and X-ray photoelectron spectroscopy (XPS), and these may also provide information about the catalytic reaction.

Specificity is an important characteristic of heterogeneous catalysis and this probably arises from the specificity of the chemisorption process. The fact that a particular gas will be chemisorbed on a particular solid under certain conditions does not mean that it will be chemisorbed on a different solid under the same conditions. Even on the same solid there may be chemisorption on one crystallographic face but not on another.

A further point to be considered is the strength of the chemisorption bond. The first molecules to be chemisorbed will go onto the sites with the largest enthalpy of desorption. It follows that the rate constants for adsorption and desorption will depend on surface coverage, with the adsorption rate decreasing as the more active sites are filled. The strength of the chemisorption bond is critical for catalytic activity: if the bond is too strong it will be difficult to break and catalytic activity will be low; if, on the other hand, the bond is too weak the surface coverage will be low and again catalytic activity will be low. There is thus an optimal bond strength between these two extremes, leading to the concept of *active sites*.

### (c) Reaction within the adsorbed layer

There are two general mechanisms that have been advanced to describe reactions within the adsorbed layer on a catalyst surface.

#### The Langmuir–Hinshelwood mechanism

This mechanism, named after its proponents, postulates that reaction takes place between two adjacent chemisorbed atoms or radicals.

For example, in the hydrogen–deuterium exchange on a metal catalyst the hydrogen and probably the deuterium are adsorbed as separate atoms of hydrogen or deuterium. The reaction might, therefore, follow the steps shown in Figure 8.12.

#### The Rideal–Eley mechanism

In this mechanism only one of the reacting species is chemisorbed, the other being either physically adsorbed or coming by direct impact from the gas phase (see, for example, Figure 8.13).

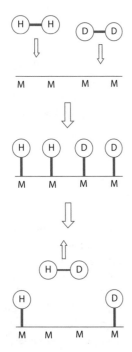

**Figure 8.12.** The Langmuir–Hinshelwood mechanism for hydrogen–deuterium exchange.

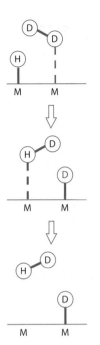

**Figure 8.13** The Rideal–Eley mechanism for the hydrogen–deuterium exchange.

$$\left.\begin{array}{c} \text{Chemisorbed radical} \\ + \\ \text{Physically absorbed} \\ \text{reactant molecule} \end{array}\right\} \Longrightarrow \left\{\begin{array}{c} \text{New chemisorbed radical} \\ + \\ \text{Physically absorbed product} \\ \text{molecule} \end{array}\right.$$

We can again look at the hydrogen–deuterium exchange as an example.

The main advantage of the Rideal–Eley mechanism is that there is no net breakage of a bond to the surface: as one bond breaks another is formed. In this way the energy change during the reaction on the surface is slight. This could be particularly important when the bond formed during the initial chemisorption is strong so that a desorption process such as that envisaged in the Langmuir–Hinshelwood mechanism is unlikely.

In the hydrogen–deuterium exchange and similar reactions there has been some controversy about the mechanism due to a report that the enthalpy of chemisorption is very high for hydrogen whereas it is known that the exchange reaction takes place rapidly even at $-183\,^\circ$C. This evidence clearly favours the Rideal–Eley mechanism, but more accurate work now indicates that the adsorption of hydrogen is reversible at $-183\,^\circ$C so current opinion now favours the Langmuir–Hinshelwood mechanism. This mechanism is also favoured by other evidence.

### Kinetics of reaction in the adsorbed layer

Rate expressions for the overall catalysed reaction can be formulated if we assume that the reaction within the adsorbed layer is the rate-determining step.

*Unimolecular reaction*  For a unimolecular reaction the rate will be determined by the concentration of the reactant at the surface. This will obviously be proportional to the fraction, $\theta(=n^\sigma/n_{\mathrm{m}}^\sigma)$, of surface covered by the reactant. Thus the rate is given by

$$v = -\frac{\mathrm{d}p_A}{\mathrm{d}t} = k\theta_A = \frac{k\alpha_A p_A}{1 + \alpha_A p_A} \tag{8.14}$$

if we assume that the Langmuir isotherm applies to the adsorption.

It follows that the reaction will be first order if $\alpha_A p_A \ll 1$, which will tend to occur at low pressures; zero order when $\alpha_A p_A \gg 1$, when pressures are high; and a fractional order in the intermediate range. The value of $\alpha_A$ will also contribute and we note that, according to Eq. (8.2), $\alpha_A$ will be large when $E_1$ is large.

*Bimolecular reaction*  For a bimolecular surface reaction in which both reactants must be chemisorbed, we can derive the rate expression as follows.

The rate expression in terms of adsorbed species is

$$v = k\theta_A \theta_B. \tag{8.15}$$

When a situation occurs in which two different gases can adsorb, a modified form of the adsorption equation is required. If the Langmuir adsorption equation can be used for both A and B, the following equations hold for their respective surface coverages (see Exercises):

$$\theta_A = \frac{\alpha_A p_A}{1 + \alpha_A p_A + \alpha_B p_B} \tag{8.16}$$

$$\theta_B = \frac{\alpha_B p_B}{1 + \alpha_A p_A + \alpha_B p_B}. \tag{8.17}$$

Thus Eq. (8.15) becomes

$$v = k \frac{\alpha_A p_A \alpha_B p_B}{(1 + \alpha_A p_A + \alpha_B p_B)^2}. \tag{8.18}$$

It is important to note the dependence on $\alpha$ and hence on the enthalpies of desorption, $E_1$ (Eq. 8.2).

If one of the reacting gases (B) is not chemisorbed, as in the Rideal–Eley mechanism, the rate expression becomes

$$v = \frac{k \alpha_A p_A p_B}{1 + \alpha_A p_A}. \tag{8.19}$$

The reaction would become first order in B (zero order in A) when $\alpha_A p_A \gg 1$ and first order with respect to both A and B (second order overall) when $\alpha_A p_A \ll 1$ (i.e. when A is only weakly adsorbed).

Other interesting limiting cases can occur with bimolecular reactions. For example, if one species (A) is adsorbed strongly or its pressure is high, while the other (B) is only adsorbed weakly, then in the denominator of Eq. (8.18) the term $\alpha_A p_A$ will dominate, and the rate equation becomes

$$v = k \frac{\alpha_B p_B}{\alpha_A p_A}. \tag{8.20}$$

That is, the very high coverage of A at the expense of B leads to a reaction with order −1 with respect to A, meaning that an increase in the pressure of the strongly adsorbed gas *retards* the reaction by reducing the ability of B molecules to adsorb.

### (d) Desorption of products
When the products of the reaction are only desorbed with difficulty (or when the surface is contaminated by a poison) it is necessary to desorb a molecule of product (poison) before the adsorption of a reactant can occur and consequently the energy barrier for adsorption is raised by an amount equal to the enthalpy of desorption of the product (poison).

## SUMMARY

A gas may be adsorbed on the surface of a solid by a process similar to condensation or by a process involving a chemical reaction. These adsorption types are known respectively as **physical adsorption** and **chemical adsorption** or **chemisorption**. The characteristics of these two adsorption types are more or less what would be expected by their similarities to condensation and chemical reaction, and as they are so different it is important to establish the type of adsorption in any experimental situation.

Methods for measuring gas adsorption are briefly discussed. They include volumetric methods, flow methods, and gravimetric methods.

Adsorption isotherms can be grouped into five types according to the shape of the isotherm plot. Equations to describe the isotherms have been developed by Langmuir, by Brunauer, Emmett, and Teller (the BET equation), and by others. The Langmuir equation describes adsorption that is limited to a single adsorbed layer (Type I) and thus describes chemisorption but does not describe the bulk of the experimental data. Most data are classified as Type II or Type IV and their shapes are well described by the BET equation. Both equations do not allow for the heterogeneity of the solid surface nor for lateral interactions between adsorbed gas molecules. They are also unsatisfactory in dealing with the thermal aspects of adsorption.

Capillary condensation can occur with porous solids at high relative gas pressures and this may lead to an apparent hysteresis in the adsorption isotherm.

In heterogeneous catalysis at least one of the reactants must be chemisorbed to the solid catalyst surface. In the Langmuir–Hinshelwood mechanism both reactants are chemisorbed whereas in the Rideal–Eley mechanism one reactant is chemisorbed but the other is either physically adsorbed or comes directly from the gas phase.

## FURTHER READING

Bowker, M. (1998). *The Basis and Applications of Heterogeneous Catalysis.* Oxford University Press, Oxford.

Hayward, D. O. and Trapnell, B. M. W. (1964). *Chemisorption.* Butterworths, London.

Ross, S. and Oliver, J. P. (1964). *On Physical Adsorption.* Interscience, New York.

Thomas, J. M. and Thomas, W. J. (1997). *Principles and Practice of Heterogeneous Catalysis.* VCH, Weinheim.

## REFERENCES

Brunauer, S., Deming, L. S., Deming, W. S., and Teller, E. (1940). *J. Am. Chem. Soc.,* **62**, 1723.

Brunauer, S., Emmett, P. H., and Teller, E. (1938). *J. Am. Chem. Soc.,* 60, 309.

Langmuir, I. (1918). *J. Am. Chem. Soc.,* 40, 1361.

Suits, C. G. (ed.) (1961). *The Collected Works of Irving Langmuir,* Pergamon Press, Oxford.

## EXERCISES

**8.1.** Use the Kelvin equation to calculate the pore radius that corresponds to capillary condensation of nitrogen at 77 K and a relative pressure of 0.75. Allow for multilayer adsorption on the pore wall by taking the thickness of the adsorbed layer on a non-porous solid to be 0.9 nm at this relative pressure. List the assumptions upon which your calculation is based. Using this result, calculate the relative pressure at which desorption should occur from this pore. Again state any assumptions used. For nitrogen at 77 K the surface tension is $8.85 \, \text{mN} \, \text{m}^{-1}$, and the molar volume is $34.7 \, \text{cm}^3 \, \text{mol}^{-1}$.

**8.2.** Derive equations (8.16) and (8.17), assuming Langmuir type adsorption for each species.

**8.3.** Derive the Langmuir isotherm equation for the situation where there is an activation energy both for adsorption ($E_a$) and for desorption ($E_d$). Deduce the appropriate expression for $\alpha$ and interpret the energy term it contains.

**8.4.** Data for the adsorption of nitrogen gas on silica, when plotted according to the linear form of the BET equation, give a slope of $319 \, \text{mol}^{-1}$ and an intercept of $3.35 \, \text{mol}^{-1}$. Calculate the surface area of the sample and, using the customary approximation, calculate the enthalpy of adsorption of the first layer. The enthalpy of vaporization of nitrogen at 77 K (the temperature of the experiment) is $5.60 \, \text{kJ} \, \text{mol}^{-1}$ and the effective cross-sectional area of a nitrogen molecule is $0.162 \, \text{nm}^2$.

**8.5.** A sample of charcoal has a monolayer capacity of 15 mg of nitrogen. In a particular experiment it was found that the sample adsorbed 2.7 mg of nitrogen at a pressure of 13.3 kPa and a temperature of 77 K (the boiling temperature of nitrogen at 101 kPa). The enthalpy of vaporization of nitrogen at 77 K is $199 \, \text{kJ} \, \text{kg}^{-1}$. Calculate the enthalpy of adsorption of the first adsorbed layer using the BET theory and appropriate approximations. State the approximations.

The following two exercises use the units that were customary in much of the earlier data.

**8.6.** The data in the table are for the adsorption of nitrogen on a 10 mg sample of mica at 90 K. Show that the data fit the Langmuir isotherm and evaluate the constants in the expression. Given that the cross-sectional area of a nitrogen molecule is $0.162 \, \text{nm}^2$, calculate the specific surface area of the mica.

| $p$/Pa | $V^\sigma$/mm$^3$ (at STP) |
|--------|----------------------------|
| 0.293  | 12.0 |
| 0.506  | 17.0 |
| 0.973  | 23.9 |
| 1.773  | 28.2 |
| 3.479  | 33.0 |

Where $V^\sigma$ is the volume of gas adsorbed at standard temperature (273.15 K) and pressure (101.32 kPa).

**8.7.** The following data are for the adsorption of ammonia on barium fluoride at 291.8 K. Using the BET equation, calculate the monolayer capacity (in cm$^3$ of gas at STP).

| $p$/kPa | $V^\sigma$/cm$^3$ |
|---|---|
| 5.26 | 9.2 |
| 8.36 | 9.8 |
| 14.40 | 10.3 |
| 21.73 | 11.0 |
| 29.19 | 11.3 |
| 36.67 | 11.8 |
| 62.12 | 12.9 |
| 80.11 | 13.4 |
| 101.97 | 14.1 |

# 9 The liquid–solid interface; colloids

## 9.1 Introduction

The removal of unwanted impurities from a liquid by passage through a column of a solid adsorbent, the separation of different solutes by column chromatography, high performance liquid chromatography, paper chromatography, are all examples of the adsorption of solutes at the liquid–solid interface and all are important processes in common use in industry. Liquid–solid interfaces are ubiquitous, from electrode surfaces to the hulls of ships.

Less obvious examples of solid–liquid interfaces are colloidal particles, dispersions of minute solid particles in a liquid. Colloids are also in daily use, from paints to blood to air pollution, and an understanding of colloids requires application of the principles developed in this chapter with a few twists that come about from their extremely small size.

## 9.2 Measurement

Adsorption on to the surface of a solid from a liquid solution is usually measured by the change in solute concentration after the solution has been brought into contact with the solid. If such measurements are to be accurate the concentration change must be appreciable and that requires the solid to have a high surface area.

Direct weighing of the adsorbed material on certain metals can be measured with the quartz crystal microbalance (QCM). A QCM consists of a quartz crystal that has been precisely cut with respect to its crystallographic axes and carries a pair of conducting electrodes. Application of an alternating electric field to the electrodes causes the crystal to vibrate. However only when the frequency of the vibration matches the natural resonance frequency of the crystal is the amplitude of the vibration appreciable. Feedback enables this natural resonance frequency to be measured. If material is deposited on the surface of the crystal the resonance frequency changes and this change is proportional to the mass of the deposited material. It is this feature that enables the QCM to be used to measure adsorption. As the electrodes cover most of the crystal surface, it is the adsorption of material onto the metal electrode that is measured. Gold is the most frequently used electrode metal, but the outer gold surfaces may be treated with other materials and their interaction with solutes can then be measured. Procedures to prevent electrical shorts and corrosion are essential for work in aqueous solutions, and it must also be remembered that changes in the density of the solution in contact with the QCM lead to changes in the response.

## 9.3 Adsorption at low solute concentration

For low solute concentrations adsorption isotherms generally have a form similar to the Type I isotherms of gas–solid adsorption and can be fitted either by the Langmuir equation (see Eq. 8.1)

$$\Gamma_B = \frac{\Gamma_B^\infty \alpha c_B}{1 + \alpha c_B} \tag{9.1}$$

or

$$\frac{c_B}{\Gamma_B} = \frac{1}{\alpha \Gamma_B^\infty} + \frac{c_B}{\Gamma_B^\infty} \tag{9.2}$$

or by the Freundlich isotherm equation,

$$\Gamma_B = \beta c_B^{1/\varphi} \tag{9.3}$$

$$\ln \Gamma_B = \ln \beta + (1/\varphi)\ln c_B \tag{9.4}$$

where $\alpha$, $\beta$, and $\varphi$ are constants (see Eq. 8.2 for the significance of $\alpha$) ($\varphi > 1$), and B indicates the solute. The adsorption isotherm shapes generated by these equations are shown in Figure 9.1.

 It is tempting to suggest that the levelling out of adsorption isotherms of the Langmuir type is due to the formation of a complete adsorbed layer on the surface of the solid, but usually this is not the case. Even when the data give a good fit to the linear form [9.2] the limiting adsorption does not usually correspond to complete coverage. The principal reason is the contribution of the solvent: it may compete for sites on the solid surface, and through its solvent power it will compete with the surface for the solute molecules.

For example, the adsorption of stearic acid on carbon black reaches different limiting values in different solvents (Figure 9.2). Note also that the acid is probably adsorbed with the acyl chain parallel to the surface and the measured adsorption values are consistent with that arrangement.

When there is strong chemisorption of the solute, the adsorption isotherm may show an almost linear adsorption with increasing solute concentration until all of the reactive sites on the solid surface are used after which there is no further chemisorption with increasing concentration.

For the adsorption of amphiphilic solutes the orientation of the adsorbed molecules depends on the nature of the surface. For charged or polar surfaces the hydrophilic part of the amphiphile will be in contact with the surface, but non-polar surfaces will attract the hydrophobic moiety.

Traube's rule is used to predict the effect of chain length on adsorption when a homologous series of long-chain substances is examined. The

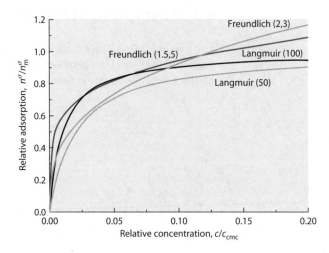

**Figure 9.1.** Adsorption isotherms calculated from the Langmuir and Freundlich equations: numbers in brackets are $\alpha c_{cmc}$ for Langmuir and $\beta/n_m$, $\varphi$ for Freundlich.

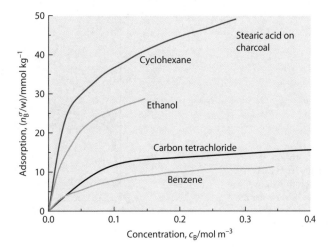

**Figure 9.2.** Adsorption isotherms for stearic acid on charcoal (Spheron 6) in various solvents. (Replotted from data of Kipling and Wright, 1962.)

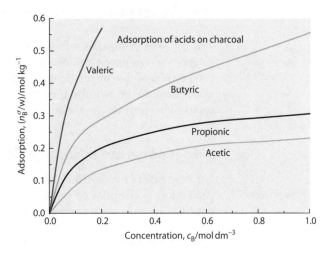

**Figure 9.3.** Adsorption of carboxylic acids from aqueous solution onto charcoal (Spheron 6) a non-polar solid, showing the effect of acyl chain length. If the adsorptions are plotted against acid activity instead of concentration all of the data lie approximately on the same curve. (Replotted from data of Hansen and Craig, 1954.)

situation is a relative one and the direction of the effect depends on the polar/non-polar characteristics of both the solvent and the solid.

For example, the adsorption of several carboxylic acids from polar solvent onto non-polar solid show increasing adsorption with increasing chain length, as increasing chain length leads to decreasing solubility and increasing interaction with the solid surface (Figure 9.3).

On the other hand, with non-polar solvent and polar solid the sequence is reversed. The interaction between the solid and the acid group is the same for all of the acids and the sequence is determined by the interaction between the solvent and the acyl chains: the longer the chain, the stronger the interaction and the lower the adsorption (Figure 9.4).

Strictly, Traube's rule describes the effect of chain length on surface tension and states that for each additional methylene group the concentration required to give a certain surface tension is reduced by a factor of 3.

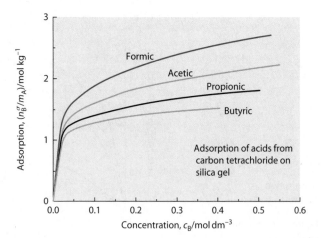

**Figure 9.4.** Adsorption of carboxylic acids from carbon tetrachloride solution onto silica gel (a polar solid), showing the effect of acyl chain length. (From data of Bartell and Fu, 1929.)

## 9.4 Adsorption at higher solute concentrations

The previous discussion refers to dilute solutions. For more concentrated solutions or mixtures of solvents, the situation is more complex as the competition of the other component for sites on the surface cannot be taken as constant as the concentration changes. The situation is described by the composite adsorption isotherm.

### 9.4.1 Composite adsorption isotherm

For a mixture of two miscible liquids, A and B, the adsorption isotherm of one component is complicated by the competitive adsorption of the other component. The apparent adsorption isotherm that is determined by measuring the change in concentration of one component (the 'solute') is often called the *composite adsorption isotherm* or sometimes the *isotherm of composition change* (see Kipling, 1965, p. 27). Thus, for example, the apparent surface excess of component B is given by:

$$n_B^\sigma = n_B^s x_A^b - n_A^s x_B^b \tag{9.5}$$

$$= n_B^s - \left(n_B^s + n_A^s\right)x_B^b \tag{9.6}$$

where $n^s$ is the total amount at the surface whereas $n^\sigma$ is the surface excess amount.

### Discussion

The two terms in Eq. (9.6) are the total amount of B on the surface and the amount that would have been there if there were no adsorption: i.e. the real amount minus the model amount, in accord with the definition of surface excess (Eq. 3.2).

Equation (9.5) indicates that the composite isotherm is related to the difference between the adsorptions of B and A and thus is easily calculated if this

**Derivation**

After adsorption equilibrium has been established the total amount of A and B in the bulk solution is

$$n^b = n_A^b + n_B^b$$

and the total amount of A and B in the system is the amounts in solution plus the amounts adsorbed

$$n^o = n_A^o + n_B^o = n_A^b + n_B^b + n_A^s + n_B^s,$$

where superscript $^o$ indicates total initial amounts, $^b$ refers to the bulk solution, and $^s$ indicates total amounts at the surface.

The total amount of B at the surface is obtained by analysis of the bulk liquid phase before and after adsorption and is therefore

$$n_B^s = n_B^o - n_B^b$$
$$= x_B^o n^o - x_B^b n^b$$
$$= x_B^o n^o - x_B^b \left[ n^o - \left( n_A^s + n_B^s \right) \right]$$

and so

$$n_B^s \left( 1 - x_B^b \right) = n_B^s x_A^b$$
$$= n^o \left( x_B^o - x_B^b \right) + n_A^s x_B^b$$

or

$$n^o \left( x_B^o - x_B^b \right) = n^o \Delta x_B$$
$$= n_B^s x_A^b - n_A^s x_B^b.$$

The quantity on the left is the adsorption or surface excess of B as determined by the standard method using concentration change. Thus

$$n_B^\sigma = n^o \Delta x_B$$
$$= n_B^s x_A^b - n_A^s x_B^b \qquad (9.5)$$
$$= n_B^s \left( 1 - x_B^b \right) - n_A^s x_B^b$$
$$= n_B^s - \left( n_B^s + n_A^s \right) x_B^b \qquad (9.6)$$

information is available. Usually, however, it is the composite isotherm that is obtained from experiment and the reverse process, deconvolution into the individual isotherms of the components, is difficult and requires additional data.

In Figure 9.5, note how negative adsorption of B can arise from the presence of the other component (A) at the surface.

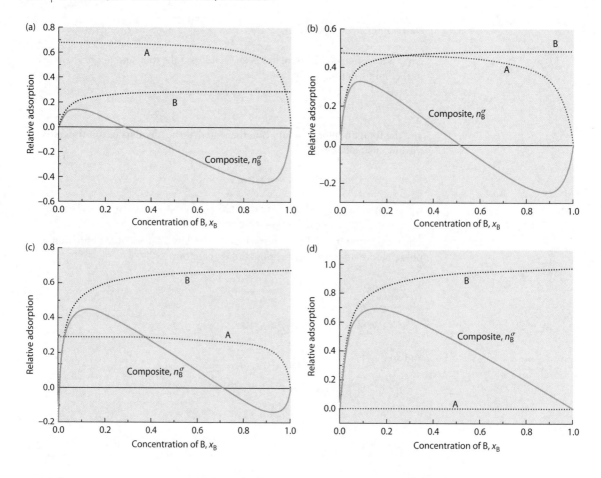

**Figure 9.5.** Computer generated composite adsorption isotherms (surface excess of B) showing how different shapes can be generated from isotherms showing the total amounts of each component at the surface.

## 9.5 Adsorption of ions and the electrical double layer

A solid surface in contact with a solution of an electrolyte usually carries an electric charge (see Section 9.9.1). We will consider a plane surface with uniform surface charge density, $\sigma$. This gives rise to an electrical potential, $\psi_o$, at the surface, and a decreasing potential, $\psi$, as we move through the liquid away from the surface, and in turn this affects the distribution of ions in the liquid.

### 9.5.1 Helmholtz model

The simplest model is that of Helmholtz: ions of charge opposite to the charge in the solid surface adsorb on the solid and completely neutralise its charge. The potential therefore falls to zero rapidly (Figure 9.6).

This model is unrealistic because it ignores the thermal motion of the ions and the solvent molecules.

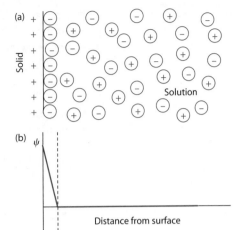

**Figure 9.6.** The Helmholtz model of the interface showing schematically (a) the distribution of ions and (b) a graph of the electrical potential.

### 9.5.2 Gouy–Chapman model

This model is based on a Boltzmann distribution of ions near the charged surface (Figure 9.7):

$$N_+ = N_+^b \exp(-z_+ e\psi/kT) \tag{9.7}$$

$$N_- = N_-^b \exp(-z_- e\psi/kT) \tag{9.8}$$

where $N_i$ is the number of $i$ ions in unit volume, $N_i^b$ is the number in unit volume of bulk solution, $z_i$ is the charge number, $e$ is the electronic charge, $k$ is the Boltzmann constant, and $T$ is temperature.

The charge density, $\rho$, at distance $x$ from the interface is:

$$\rho = \sum N_i z_i e = \sum \left[ N_i^b z_i e \exp(-z_i e\psi/kT) \right]. \tag{9.9}$$

Note that for a symmetrical electrolyte (where $z_+ = -z_- = z$) Eq. (9.9) can be written:

$$\rho = 2N^b z e \sinh(-ze\psi/kT). \tag{9.10}$$

Combination of (9.9) and (9.10) with the Poisson equation for a plane interface:

$$\frac{d^2\psi}{dx^2} = \frac{-\rho}{\varepsilon} \tag{9.11}$$

where $\varepsilon$ is the permittivity of the liquid, gives the relevant forms of the Poisson–Boltzmann equation:

$$\frac{d^2\psi}{dx^2} = -\sum \left[ (N_i^b z_i e/\varepsilon) \exp(z_i e\psi/kT) \right] \tag{9.12}$$

and for a symmetrical electrolyte:

$$\frac{d^2\psi}{dx^2} = -2 \left[ (N^b z e/\varepsilon) \sinh(-ze\psi/kT) \right]. \tag{9.13}$$

If $|ze\psi| \ll kT$ (known as the Debye–Hückel approximation) the series expansions of the exponentials in (9.12) and (9.13) can be truncated after the second term, giving:

$$\frac{d^2\psi}{dx^2} = -\sum (N_i^b z_i e/\varepsilon) + \sum (N_i^b z_i e/\varepsilon)(z_i e\psi/kT). \tag{9.14}$$

However, because of electrical neutrality in the bulk of the solution, the first summation term on the right of (9.14) is zero and we have:

$$\begin{aligned}
\frac{d^2\psi}{dx^2} &= \sum (N_i^b z_i e/\varepsilon)(z_i e\psi/kT) \\
&= (e^2\psi/\varepsilon kT) \sum N_i^b z_i^2 \\
&= \kappa^2\psi
\end{aligned} \tag{9.15}$$

where $\kappa$, defined by

$$\kappa^2 = (2e^2 N_A/\varepsilon kT)I, \tag{9.16}$$

is independent of $x$, and where $N_A$ is the Avogadro constant and $I$ is the ionic strength (in mol m$^{-3}$) defined by:

$$I = \frac{1}{2N_A} \sum_i N_i z_i^2. \tag{9.17}$$

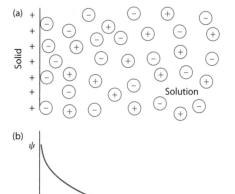

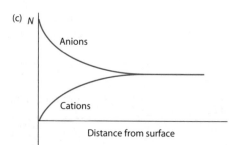

**Figure 9.7.** The Gouy–Chapman model of the interface showing schematically: (a) the distribution of ions, (b) a graph of the electrical potential, and (c) the concentrations of ions.

The Debye length

The quantity $\kappa^{-1}$ has the dimensions of length and is called the Debye length. It is a measure of the thickness of the diffuse layer and equals the thickness of the equivalent parallel plate condenser. Note that as $I$ increases $\kappa^{-1}$ decreases, showing the contraction of the diffuse layer.

The solution of equation (9.15) is:

$$\psi = \psi_o \exp(-\kappa x) \tag{9.18}$$

where $\psi_o$ is the potential at the beginning of the diffuse layer. This is situated at the surface of the solid in the Gouy–Chapman model, but not in the Stern modification that will be described later.

Calculation shows that (9.18), which is based on the Debye–Hückel approximation, is valid only for $\psi_o < 25\,\text{mV}$, which is lower than many observed potentials. Nevertheless (9.18) is sufficient to show, in a semi-quantitative way, the variation of $\psi$ with distance, $x$, from the surface and with ionic strength, $I$.

The Gouy–Chapman model describes a diffuse layer of ions in which the charge on the surface is progressively neutralised by an excess of oppositely charged ions as we move away from the surface (Figure 9.7).

### 9.5.3 The Stern model

A difficulty with the Gouy–Chapman model arises from the treatment of ions as point charges. As a consequence, for conditions of high $\psi_o$ and high ionic strength the calculated concentration of real ions, occupying volume, near the surface is greater than can be accommodated in the volume available. A simple means for treating this problem is due to Stern, who proposed that the size of the ions should be considered for only the first layer of adsorbed ions, with the ions further away being treated as point charges as in the Gouy–Chapman theory. The model thus proposes two regions: the Stern layer immediately adjacent to the surface where ion size is important; and outside this a diffuse layer where the Gouy–Chapman treatment still applies (Figure 9.8).

Equation (9.18) then becomes:

$$\psi = \psi_h \exp\{-\kappa(x - x_h)\} \tag{9.19}$$

where $x_h$ is the thickness of the first adsorbed layer of ions (the Stern or Helmholtz layer), and $\psi_h$ is the potential at $x_h$.

The electrical double layer makes an important contribution to the stability of colloidal dispersions and provides an explanation for the effect of added electrolytes on the stability of colloids. An important tool in the examination of the electrical double layer is the set of phenomena known as electrokinetics.

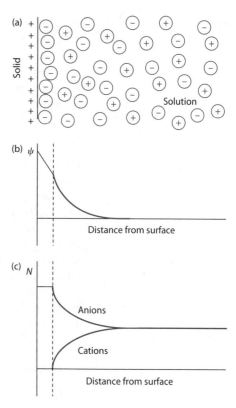

**Figure 9.8.** The Stern model of the interface showing schematically: (a) the distribution of ions, (b) a graph of the electrical potential, and (c) the concentrations of ions. The Stern layer is indicated by the dashed line.

## 9.6 Electrokinetics

Because of the difference in charge between the diffuse layer and the solid surface, movement of one relative to the other will cause charge separation and hence generate a potential difference, or alternatively, application of an electrical potential will cause movement of one relative to the other. Thus there are four electrokinetic phenomena as shown in Table 9.1.

In a rather simplified picture of the electrokinetic phenomena, the relative movement of the solid surface and the liquid occurs at a surface of shear. The potential at this surface of shear is known as the zeta ($\zeta$) potential and its value can be determined by measurement of one of the electrokinetic phenomena.

The importance of the $\zeta$ potential comes from the postulate that the surface of shear is almost identical with the surface at the outside of the Stern layer. The $\zeta$ potential thus gives a measure of the potential at the beginning of the diffuse layer and hence corresponds to $\psi_o$ in Eq. (9.18) or to $\psi_h$ in Eq. (9.19). It also gives an indication of the extent to which ions from the solution are adsorbed into the Stern layer. When the solid surface is able to ionize, as with oxides and proteins, for example, the $\zeta$ potential is a measure of the extent of ionization (Hunter, 1993, p. 255).

**Table 9.1.** The electrokinetic phenomena.

| Stationary phase $\rightarrow$ | Solid | Liquid |
|---|---|---|
| Moving phase $\rightarrow$ | Liquid | Small particles |
| Movement by electric field | Electro-osmosis | Electrophoresis |
| Mechanical movement | Streaming potential | Sedimentation potential |

Two other terms need to be mentioned. The isoelectric point is the pH at which the $\zeta$ potential is zero. It is determined by a pH titration: measuring the $\zeta$ potential as a function of pH. The point of zero charge is the pH at which the positive and negative charges of a zwitterionic surface are balanced. It is the same as the isoelectric point if there is no specific adsorption of ions onto the surface.

## 9.7 Colloidal dispersions

Colloids are dispersions of microscopic or submicroscopic particles in a fluid medium. If the dispersion medium or continuous phase is a gas the term aerosol is applied; if the dispersion medium is a liquid the terms sol, or hydrosol (when the medium is water) are used. The diameters of colloidal particles lie between 1 nm and about 5 µm. Mostly the particles are solids, but liquid droplets can also form colloidal dispersions as in the microemulsions (see Section 6.4).

One important consequence of the small size of colloidal particles is that collectively they have a very large area of interface with the dispersion medium. The properties of this interface are therefore very important in the properties of colloids.

Colloidal dispersions are found in many industrial areas, such as in the manufacture of certain foods, cosmetics, and paints, in oilfields, and in many biological systems. Much air and water pollution is also colloidal in nature.

### 9.7.1 Lyophilic and lyophobic colloids

The sols are usually divided into two classes according to the interaction between the particles and the dispersion medium.

Lyophilic (*solvent loving*) colloids are characterized by the ease with which the dry colloidal material can be dispersed in the solvent or dispersion medium. Often the lyophilic colloids are simply large molecules, such as proteins and certain polymers, where single molecules fall into the colloidal size range. Thus in essence, many lyophilic colloids are solutions of large molecules. The lyophilic colloids are thermodynamically stable and are therefore very difficult to flocculate.

Lyophobic (*solvent hating*) colloids, on the other hand, are difficult to disperse from the dry material, may be easily coagulated, and are thermodynamically unstable. Given time, they will coagulate spontaneously, although the time scale may be very long.

## 9.7.2 Preparation of colloids

Methods for preparing colloidal dispersions can be classified as dispersion techniques or aggregation techniques.

In dispersion techniques the bulk material is ground down to the required size range in a colloid mill or by using a high speed stirrer. The presence of the dispersion medium is usually desirable. Various industrial designs have been developed (Hunter, 2001; Alexander and Johnson, 1950, p. 557).

Aggregation methods usually involve precipitation from a supersaturated solution. The process can be divided into three steps: formation of the supersaturated solution by chemical reaction, sudden cooling, or removal of a solvent; formation of nuclei by homogeneous nucleation or use of pre-existing nuclei; growth of particles by precipitation on the nuclei. Removal of solvent involves the dissolution of the substance in a water-miscible solvent such as ethanol, then adding this solution to boiling water. Further boiling removes the solvent leaving a supersaturated solution of the substance in water.

Electric arc methods involve dispersion of the metal of the electrodes in the electric arc and subsequent aggregation of the vapour.

### Dialysis

Dialysis is a procedure for removing excess ions and other unwanted solutes from a sol. The dispersion is placed in a bag formed from a membrane with pores large enough to permit the passage of the unwanted solutes but small enough to retain the colloidal particles. The bag is placed in a wash solution and the unwanted solute molecules move through the membrane by diffusion. As the rate of diffusion depends on the concentration difference, it is important to replenish the wash solution frequently or to use a flow system.

### Monodisperse colloids

By careful preparation it is possible to produce colloids with all the particles of approximately the same size. The principles of the technique have been developed by LaMer and his associates (e.g. Zeiser and La Mer, 1948). In essence, very dilute solutions are used so that the reaction occurs slowly, forming molecularly dispersed product whose concentration increases through the supersaturation stage to a value at which stable nuclei form by homogeneous nucleation. This initial production of nuclei causes the concentration of molecularly dispersed product to fall below the value for self-nucleation so that as the reaction slowly continues no new nuclei are formed and the product is deposited on the previously formed nuclei. As this process is slow (because of the low concentration) all of the nuclei grow to nearly the same size.

Monodisperse sulfur sols

A monodisperse sulfur sol can be prepared as follows. Firstly it is essential that all vessels be scrupulously clean and that reagents be purified to remove any particulate matter that could act as heterogeneous nuclei for the precipitation. To a dilute solution of $Na_2S_2O_3$ ($1\,dm^3$ of $0.001\,mol\,dm^{-3}$) at $25\,°C$ is added rapidly, with stirring, sufficient concentrated acid (HCl or $H_2SO_4$) to give $0.003\,mol\,dm^{-3}$ of $H^+$. The reaction proceeds slowly and after about an hour a Tyndall beam can be detected and higher order Tyndall spectra begin to develop. The reaction can be stopped by titrating the unreacted thiosulfate with an iodine solution (about $3.5\,cm^3$ of $0.1\,mol\,dm^{-3}\,I_2$) (La Mer et al. 1946, 1950).

$$S_2O_3^{2-} + H^+ \longrightarrow HSO_3^- + S\downarrow$$

$$2S_2O_3^{2-} + I_2 \longrightarrow 2I^- + S_4O_6^{2-}.$$

The sulfur particles are reported to be amorphous and spherical and their size can be determined from the higher order Tyndall spectra (see Section 9.8.2).

## 9.8 The properties of colloids

### 9.8.1 Brownian motion

In the nineteenth century Robert Brown observed through a microscope the jiggling motion of pollen grains suspended in water, now known as Brownian motion. This motion is caused by impacts of the solvent molecules on the pollen grains and although colloidal particles are generally too small to be observed through a microscope it is to be expected that they would be subjected to the same impacts and perform similar jiggling motions.

Brownian motion is an important aspect of colloidal behaviour at it is responsible for the diffusion of colloidal particles, for collisions between colloidal particles, and for assisting in maintaining colloidal particles in suspension. Unless the particles have the same density as the dispersion medium they will tend, due to the gravitational force, to sink to the bottom of the vessel or to rise to the upper surface (known as creaming), processes that are retarded by Brownian motion.

### 9.8.2 Light scattering

While larger colloidal particles (about $1\,\mu m$ diameter) can be directly observed through an optical microscope, smaller particles cannot be seen. However, if the dispersion is illuminated from the side, the path of the light beam can be observed because of the light scattered by the particles. This is known as the Tyndall effect.

Use is made of the Tyndall effect in an instrument known as an ultramicroscope. Light is focused on the colloidal dispersion and the light scattered at right angles is examined through a microscope. Even when the particles themselves are too small to be observed, the light that they scatter can be seen through the microscope as pin-points of light. This instrument can be used to observe the movement of colloidal particles in an electric field and thus measure their electrophoretic mobility from which, with suitable calibration, the $\zeta$-potential can be calculated.

Light may be scattered in all directions and a study of the angular distribution of the scattered light can provide information about particle size. The scattering patterns may be divided into two kinds: **Rayleigh scattering** occurs if the particles are much smaller than the wavelength of the light, but with larger particles there can be interference of light scattered from different parts of the same particle and a much more complex pattern ensues, as described by Mie. Most colloids fall into the second category so that **Mie scattering** is the usual pattern.

Of special interest is the scattering from monodisperse colloids. With white light the different wavelengths generate their greatest scattering intensities at different angles so that scanning though the scattering angles one observes a spectrum of colours. This is sometimes called the **higher order Tyndall spectrum**. By measuring the scattering angle for the maximum intensity of one colour (often red or green) the particle size can be calculated.

This topic is too complex for detailed discussion here, but further information can be found in Scheludko (1966) and Hunter (2001).

---

### Sol–gel technology: new materials from colloids

Ceramics have been known and used since early civilization, but improvements in synthesis and processing continue to be made. Modern ceramics are the basis of advanced materials for a wide range of demanding applications. Many of these are produced by chemical synthetic routes, and one such method is sol–gel processing. In fact the term sol–gel is used to describe several types of processes which can be used to form a wide variety of materials, and in most of these, colloid science plays an integral role.

The sol–gel process involves the formation of inorganic materials from a starting material, such as a metal alkoxide, being dispersed as a colloidal dispersion (sol), and the subsequent gelation of the sol to form a continuous network. The basic chemical steps involved are:

metal alkoxide solution → sol (via hydrolysis, polymerization)

→ gel (via gelation by removal of solvent)

→ further processing (e.g. heating, spinning)

Typical starting materials are TMOS (tetramethyl orthosilicate) or TEOS (tetraethyl orthosilicate). The great advantage of these techniques is that they allow the production of advanced ceramics at, or near to, room temperature. They can also be used to produce materials in a number of forms, such as dense ceramics, thin films, low density materials (known as aerogels) and even ceramic fibres.

### 9.8.3 Measurement of particle size

The size of colloidal particles can be measured by light scattering as discussed above. Some other methods will be described here. A more extensive and more detailed description of sizing methods is given by Hunter (2001).

#### Dynamic light scattering

Quasi-elastic light scattering (QELS) or photon correlation spectroscopy occurs when the frequency of the light scattered by a particle is different from the frequency of the incident light due to movement of the particle. This Doppler broadening gives a value for the diffusion coefficient from which the particle size can be estimated.

#### The electron microscope

The electron microscope is useful for viewing colloidal particles but there are some limitations. Both transmission (TEM) and scanning (SEM) electron microscopes may be used. However with both instruments the dispersion medium must be removed as the sample must be examined in a vacuum, so there are cases where the sample preparation will change the particles. Also the particles must be able to withstand the intense electron beam without damage.

There are some ancillary techniques for examining samples that have been deposited on an electron microscope grid and the dispersion medium removed. They include replication which is useful when the sample is sensitive to the electron beam or to the vacuum. It involves casting a replica of the surface using a material (such as carbon) that will withstand the electron beam. The image may be enhanced by shadow casting where the sample or its replica is exposed to a beam of metal atoms (usually gold or platinum) at a low angle in vacuum. The resultant pattern is normally stable in the electron beam so this is a very important technique for samples that would be damaged by direct exposure to the beam. Measurement of the length of a shadow and knowledge of the angle of shadowing give the height of the particle.

Cryoelectron microscopy has the advantage over many other techniques of sample preparation in that it is not necessary to remove the dispersion medium. It is particularly useful when the dispersion medium is water. A drop of the colloidal dispersion is placed on a coated electron microscope grid, blotted, and then plunged rapidly into liquid propane. Vitrification of the water is rapid and preserves the structure of the sample because crystallization, with the associated expansion, is avoided. The grid can then be transferred into liquid nitrogen for storage and later examined on the cold stage of a special electron microscope (Battersby *et al.*, 1994).

#### The Coulter counter

In this instrument the colloidal solution is forced to flow through a small orifice with an electrode on either side. The solution must contain a small amount of electrolyte (about 5%) so that when a particle passes through the orifice there is a pulse in the conductance between the two electrodes. The amplitude of each pulse is proportional to the volume of the particle moving through the orifice so a pulse height analyser gives the particle size distribution of the colloid.

### Electroacoustics

If a colloidal dispersion is subjected to high frequency sound the particles move back and forth in response, but the ions in the double layer move to a greater extent and more quickly. This disparity generates small dipoles and the aggregate of these dipoles oscillating in unison produces an electrical signal that can be detected by electrodes placed in the dispersion at the peak and trough of the sound wave.

The reverse phenomenon where an alternating electric field is applied and a sound wave is generated and detected now forms the basis of a commercial instrument. There is a phase lag due to the inertia of the particles which depends on the size of the particles. Hence measurements of the phase lag over a range of electric frequencies yields the size distribution, and it is also possible to determine the zeta potential in the same measurement. One advantage of the method is that it works in concentrated dispersions where optical methods would not be suitable.

### Field flow fractionation

In these methods the colloidal dispersion flows through a thin channel while subjected to a field at right angles to the flow. The field may be gravitational (usually centrifugal), electrical, magnetic, or thermal. The effect depends on the flow profile in the channel: flow near the walls, particularly the wall towards which the field is directed, is slower than flow at the centre of the channel, so particles that are less affected by the field are eluted faster than particles that are forced towards the wall. The system works in a manner similar to that of chromatography: a small sample is injected into the stream and monitored at the exit port. Good separations of particles according to size can be achieved.

## 9.8.4 Measurement of particle charge

The actual charge on the surface of a colloid can often be determined by titration or by ion exchange. The details of the titration depend on the nature of the particles and further details can be found in Hunter (2001).

Generally it is the $\zeta$-potential that is of greater interest as this has more relevance to the stability of the dispersion.

A simple and semiquantitative apparatus for measuring electrophoretic movement and hence estimating the $\zeta$-potential is shown in Figure 9.9. It consists of a U-tube fitted with a wide-bore tap at the base of the U connected in turn to a reservoir. The reservoir and connecting tube up to and including the tap are filled with the colloidal suspension and the tap is closed. Clear solvent is added to the empty U-tube and then, with the hydrostatic pressures carefully balanced the tap is opened. Raising the reservoir then allows the

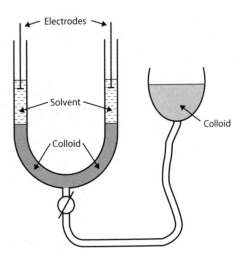

**Figure 9.9.** Simple Burton tube for the measurement of electrophoresis. Application of an electrical potential difference (usually about 200 V DC) to the electrodes causes the solvent–colloid boundaries to move towards the electrode with the opposite charge to that of the colloidal particles.

colloid to flow slowly into the U-tube so that it does not mix with the solvent but lifts it up the arms of the tube. If this operation is performed carefully there will be a sharp interface between the colloidal solution and the overlying solvent. Insertion of an electrode (usually Ag|AgCl) into each arm of the U-tube allows an electrical potential to be applied. The resulting movement of the solvent–colloid boundaries can then be measured and the electrophoretic mobility calculated.

A refinement of this simple apparatus was developed by Tiselius (1937). The U-tube is divided into sections with ground glass plates at top and bottom so that each section can be moved horizontally relative to the others. Thus the lower sections can be filled with colloid, pushed aside, and the upper sections flushed and filled with solvent. When the sections are realigned there is a sharp solvent–colloid boundary which can be brought into view (away from the junction) by adding extra solvent to one side. Schlieren optics are used to detect the boundaries and measure their movement when a potential difference is applied to the electrodes. Because the schlieren optics detects changes in refractive index, boundaries that are not necessarily visible to the eye can be monitored. For this reason the apparatus is used extensively for the examination of such materials as protein dispersions.

As mentioned earlier, the ultramicroscope can be used to observe the movement of colloidal particles even when the particles themselves are too small to be directly observed by an optical microscope. For electrophoresis measurements the colloid is usually held in a wide-bore capillary tube of known dimensions and an electric field is applied through electrodes placed at each end of the tube. The colloid particles move by electrophoresis in response to the electric field, but the liquid near the tube walls moves in the opposite direction by electro-osmosis and flows back down the centre of the tube. Thus the observed particle movement varies across the tube but there is a particular distance from the wall where the electro-osmotic flow is zero so measurements at that position (found by focusing the microscope on the wall and then moving it the required distance) give the true electrophoretic mobility. Several commercial models are available. For further detail see Hunter (1993, p. 241).

## 9.9 Coagulation of lyophobic colloids by electrolytes

The terms *flocculation* and *coagulation* tend to be used interchangeably in colloid science, but sometimes a distinction is made: flocculation being a loose aggregation of colloidal particles whereas coagulation involves a closer aggregation or even a merging of the particles.

The addition of an inorganic salt has little effect on the stability of lyophilic colloids unless the concentration is extremely high, when salting out may occur. In contrast, the addition of salts in quite low concentrations to a lyophobic sol often causes flocculation or coagulation.

### 9.9.1 Surface charge

Lyophobic colloids usually carry a surface charge which can originate in a variety of ways. In some cases the particles contain ionizable groups, such as $-COOH$ or $-NH_2$, which can ionise or attract ions from the solution to generate a net charge on the particle. Metal oxides can form $M-O^-$ or $M-OH_2^+$ at the surface depending on the pH. Inorganic colloids (the classic example is AgI) possess a small but significant solubility product so that addition of the relevant cation or anion (such as $Ag^+$ or $I^-$) to the dispersion medium causes an imbalance of the ions on the surface of the particles. Ions which determine the charge on the particle surface by such mechanisms are called potential determining ions.

### 9.9.2 The Schulze–Hardy rule

Indifferent ions which have no specific interaction with the surface can nevertheless affect the behaviour of the colloid. Addition of an indifferent electrolyte to a lyophobic colloid can decrease its stability, primarily due to a contraction of the double layer (Section 9.9.3), and it is possible to determine a concentration that leads to rapid coagulation. This is known as the critical

---

**Silver iodide sol**

For the silver iodide sol, the point of zero charge is when there are equal numbers of $Ag^+$ and $I^-$ ions on the surface. However, with equal concentrations of $Ag^+$ and $I^-$ in the solution the surface charge is negative which implies that the $I^-$ ions have a greater affinity for the surface. Thus at the point of zero charge there must be a higher concentration of $Ag^+$ ions than $I^-$ ions in the solution.

coagulation concentration. It is determined by adding different amounts of the electrolyte to a set of test tubes containing the colloid and observing the concentration above which coagulated material is formed within a predetermined time. Note that the critical concentration depends to some extent on the waiting time, so standardization is essential for comparative measurements.

The effectiveness of the salt in causing flocculation depends on the charge of the ion of opposite sign to the charge on the sol particles. Trivalent ions are much more effective than divalent ions, which in turn are more effective than monovalent ions with the relative coagulating powers being very roughly in the ratios $1000 : 100 : 1$. These observations are sometimes called the Schulze–Hardy rule. For example, critical coagulation concentrations for the negatively charged $As_2S_3$ sol include: NaCl, 51; $MgCl_2$, 0.72; $AlCl_3$, 0.093 mol m$^{-3}$.

There is also a relatively small effect of ion hydration on flocculation. If we take a series of flocculating ions with the same charge and the same counterion, their ability to produce flocculation depends on the extent of hydration. For example, for a negatively charged *hydrophilic* colloid, flocculating power increases with the extent of hydration, so for monovalent cations we have the sequence $Li^+ > Na^+ > K^+ > NH_4^+ > Rb^+ > Cs^+$, as the high concentrations of these ions required for flocculation tend to dehydrate the colloidal particles: a salting out effect. For a negatively charged *hydrophobic* colloid, however, the sequence is reversed: $Li^+ < Na^+ < K^+ < NH_4^+ < Rb^+ < Cs^+$, as the size of the hydrated ion restricts its ability to approach the particle surface and enter the Stern layer. Similar series can be described for anions and for divalent ions. Collectively they are known as the lyotropic or Hofmeister series.

Because of the importance of lyophobic colloids, a great deal of attention has been given to finding an explanation for their (meta) stability and for the effect of electrolytes. This has led to detailed theoretical studies of the forces between colloidal particles and, more recently, to actual measurements of such forces. The theory, however, is developed in terms of potential energies rather than forces as the treatment is simpler and the concept of energy barriers is more familiar.

### 9.9.3 Stability of colloidal dispersions – DLVO theory

Lyophobic colloidal suspensions are thermodynamically unstable, i.e. there is a free energy decrease when the individual particles gather into a bulk phase, or into clusters of solid particles. However colloidal suspensions may appear to be stable because there is an energy barrier which must be overcome for the particles to *coagulate* or *flocculate*. Such a suspension could be described as *metastable*.

The forces involved are:

- attractive – the long range van der Waals force or Hamaker force arising from induced-dipole–induced-dipole interactions between the assemblies of particles in each particle;

- repulsive – due to interaction of similarly charged double layers surrounding particles;
- Born repulsion – is a very short range force arising from the mutual repulsion of the electrons associated with the atoms of each particle; but is not usually considered in stability theories.

The overall potential energy of interaction ($V_t$) is the sum of the attractive and repulsive potentials:

$$V_t = V_a + V_r. \tag{9.20}$$

However, it is the change in the overall potential energy as two particles approach that determines whether they experience an attractive or a repulsive force. If the energy decreases as they approach, they experience an attraction; if the energy increases, they experience a repulsion.

It is worth noting that the charge on an isolated particle in a dispersion is completely neutralized by its electrical double layer. Thus the repulsive electrical force is only experienced when two particles approach closely enough for their double layers to overlap. Simpler ideas of a pure electrostatic repulsion between like-charged particles are not appropriate.

There were a number of early contributions to the modern theory describing the stability of lyophobic colloidal dispersions. Hamaker (1937) developed an expression for the attractive potential, and the full theory came independently from the work of Derjaguin and Landau (1941) and Verwey and Overbeek (1948), and is now known as DLVO theory.

### Attractive potential energy

The potential energy of attraction for similar spherical particles of radius $r$ in a vacuum, with centres separated by a distance $(s + 2r)$ is given by

$$V_a = \frac{-Hr}{12s} \tag{9.21}$$

where $H$ is the *Hamaker constant*, and the separation, $s$, between the particle surfaces is small relative to their radii. $H$ is always positive if the two particles are of the same material, so this potential is always negative, i.e. always attractive. In a liquid medium, $H$ is replaced by an effective Hamaker constant given by

$$H = \left( \sqrt{H_p} - \sqrt{H_m} \right)^2 \tag{9.22}$$

where $H_p$ and $H_m$ are the Hamaker constants for the particles and the dispersion medium respectively.

### Repulsive potential energy

The repulsive potential arises from the interpenetration of the diffuse double layers surrounding the particles. The theory is based on equilibrium situations but if the particles are approaching one another the adjustment of the double layers may lag behind the movement so that equilibrium is not immediately achieved. In the DLVO theory the surface potential is taken to be constant throughout the approach and in the equations given below the two surfaces are identical. The equilibrium repulsion depends on the size and shape of the particles, the distance between them, the surface potential, the ionic strength and the dielectric constant of the dispersing liquid. Approximate solutions are given by Hunter (2001):

$$V_r = 2\pi\varepsilon r\psi_o^2 \exp(-\kappa s) \qquad \text{for } \kappa r \ll 1$$
$$V_r = 2\pi\varepsilon r\psi_o^2 \ln(1 + \exp(-\kappa s)) \text{for } \kappa r \gg 1$$

(9.23)

where $\psi_o$ is the potential at the surface of the particles, $s$ is the distance of closest approach, and $\kappa$ is the inverse of the Debye length. Note that $\kappa r = r/(1/\kappa)$ is the ratio of the particle size to the double layer thickness.

The contraction of the double layer interaction with increasing ionic strength is shown clearly in Figure 9.10.

### Overall potential energy

The overall potential energy of interaction can be plotted as a function of distance between two particles by adding the van der Waals–Hamaker attraction and the double-layer repulsion as in Figure 9.11.

The shape of the potential curve depends on the size of the particles, the electrolyte concentration, and the surface potential, $\psi_o$. For a stable dispersion, an energy barrier of ~15 $kT$ ($= 62 \times 10^{-21}$ J at 298 K) or greater is needed.

Influence of r

Equations (9.21) and (9.23) show that the interaction energies are proportional to the particle radius. Thus for a particular set of conditions the energy barrier to coagulation is proportional to the radius.

Influence of electrolyte concentration

An increase in the electrolyte concentration leads to a compression of the double layer ($\kappa$ increases) (see Figure 9.10) and so the energy barrier to coagulation decreases or disappears.

**Figure 9.10.** Effect of increasing ionic strength ($I$) on the double layer interaction between two identical spherical particles, calculated from Eq. (9.23) for $\kappa r \gg 1$.

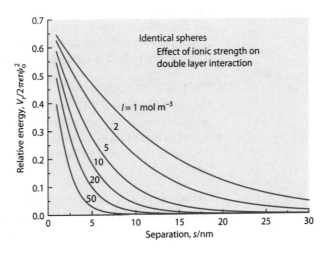

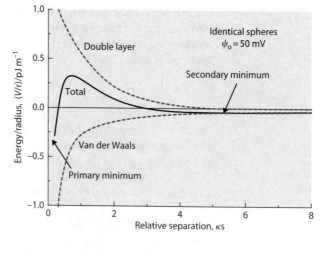

**Figure 9.11.** Addition of the contributions from the van der Waals attraction and the double-layer repulsion gives the overall potential energy of interaction between two particles. Note that the depth of the primary minimum is limited by the Born repulsion at short separations, although this contribution has not been shown.

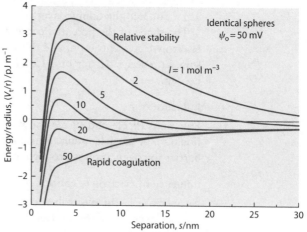

**Figure 9.12.** Potential energy curves for the approach of two identical spherical particles showing the effect of ionic strength (values shown on the graph). Note that in these calculations the potential at the surface, $\psi_o$, has been held constant.

As the ionic strength is raised the double layer contracts, weakening the repulsion and lowering the energy barrier. At a sufficiently high ionic strength (for example, the curve for $50 \, \text{mol m}^{-3}$ in Figure 9.12) there is no energy barrier, but a continuing attractive potential as the particles approach. Thus every near approach leads to coagulation.

### Effect of surface potential

The curves in Figure 9.12 were calculated with the potential at the surface ($\psi_o$) held constant. In practice, the addition of electrolyte would tend to lower $\psi_o$ and coagulation would then be expected at a lower ionic strength. The effect of changes in the surface potential at constant ionic strength is shown in Figure 9.13.

### Overall effect of electrolyte addition

In Figure 9.12 and Figure 9.13 the effects of added electrolyte in reducing the extent of the double layer and lowering the potential at the surface have been

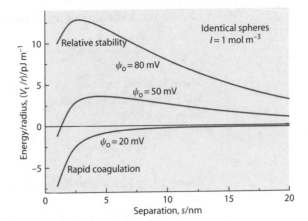

**Figure 9.13.** The effect of changes in surface potential in the total interaction potential energy. Note that the low surface potential of 20 mV leads to rapid coagulation at this low ionic strength.

artificially separated. However, in practice both effects would occur simultaneously and combine to reduce the double layer repulsion and consequently the stability of the dispersion.

It is worth noting that ionic strength depends on the squares of the ionic charges (Eq. 9.17) and through this relationship the DLVO theory provides a partial explanation for the Schulze–Hardy rule relating to the effect of charge number on the coagulation values for different electrolytes. The other factor in the Schulze–Hardy rule is probably the effect of added ions on the surface potential, as in Figure 9.13.

The secondary minimum

At relatively large separations the van der Waals attraction is greater than the double layer repulsion giving rise to a secondary minimum as shown in Figure 9.11 and Figure 9.12. Particles may therefore aggregate with a large distance between them, a process sometimes called *flocculation*. Since the secondary minimum is quite shallow, flocculation of this type is easily reversible, and the particles can be separated by agitation.

### 9.9.4 Kinetics of coagulation

The stability of a dispersion can be measured by determining the rate of change in the number of particles, $N_p$, during the early stages of aggregation. In the absence of an energy barrier, the rate of coagulation is determined by the rate of diffusion-controlled collisions. This is described by the *von Smoluchowski equation*:

$$-\frac{dN_p}{dt} = 8\pi D r N_p^2 \tag{9.24}$$

where $D$ is the diffusion coefficient. Using the Einstein equation for the diffusion coefficient:

$$D = \frac{kT}{6\pi\eta r} \tag{9.25}$$

where $\eta$ is the viscosity of the dispersion medium, we have:

$$-\frac{dN_p}{dt} = \frac{4kT}{3\eta}N_p^2 = k_o N_p^2 \qquad (9.26)$$

where $k_o$ is the rate constant for diffusion-controlled coalescence. If, on the other hand, there is an energy barrier, $V_{max}$, to coagulation, this equation becomes

$$-\frac{dN_p}{dt} = k_o N_p^2 \alpha = \left(\frac{4kT}{3\eta}\right)N_p^2 \exp\left(\frac{-V_{max}}{kT}\right) \qquad (9.27)$$

where $\alpha$ is a factor introduced by Smoluchowski to allow for unsuccessful collisions.

Overall, according to DLVO theory the stability of a dispersion is increased by:

- an increase in the particle radius;
- an increase in the surface potential;
- a decrease in the effective Hamaker constant;
- a decrease in the ionic strength of the dispersing liquid;
- a decrease in temperature.

### 9.9.5 Limitations of DLVO theory

DLVO theory is very useful for predicting effects on colloid stability which are electrical in origin, such as the effect of added inorganic ions or added surfactant where the main result is a change in the surface potential of the particles. There are many cases, however, where the effect of added surfactant arises from other factors, such as steric effects (e.g. polymer surfactants), or the adsorption of surfactant on to the particles with a resulting change in the contact angle. In these cases DLVO theory must be used with caution.

At close distances of approach the solvation of the solid surfaces can have a significant effect on the interaction force that is not considered in the DLVO theory. Furthermore, the packing of solvent molecules in the space between the two particles can cause pronounced oscillations in the interaction force (see Section 9.10.1).

### 9.9.6 The measurement of interparticle forces

The development of the DLVO theory has generated considerable interest in the measurement of the forces between colloidal particles. Early measurements suffered from a variety of experimental difficulties, but in recent years techniques have been developed that have enabled the forces between two surfaces to be measured as a function of their separation. We will briefly describe two techniques and for further details of these and other techniques refer to Hunter (1993), Israelachvili (1991), and Ducker (1991).

The direct measurement of surface forces with the apparatus of Israelachvili uses curved mica sheets because of their smoothness. The thin mica sheets are silvered on the back faces and glued to polished silica surfaces arranged as crossed cylinders. For force measurements this geometry is equivalent to a spherical particle approaching a flat surface. An arrangement of springs is used to move the lower surface towards the fixed upper surface and the separation between the sheets is measured by interferometry using the fringe pattern formed by reflections from the silver backing. The lower surface is mounted on a weak cantilever spring so that when a repulsion between the surfaces is experienced the spring bends and the separation between the surfaces is greater than that expected from earlier calibration. Knowledge of the spring constant enables the force to be calculated. It is therefore possible to measure the forces between the surfaces as a function of their separation.

Measurements have been made in a variety of aqueous solutions and have been found to follow the exact double-layer theory. Hydration of ions on the surface has been found to provide a very short range repulsion in certain conditions and, in some cases, has shown a superimposed oscillation with a period that corresponds to the size of a water molecule.

Interparticle forces have also been measured by a modification of the atomic force microscope (AFM). General details of the technique are given in Section 7.3.2. For measuring interparticle forces the scanning tip is replaced by a small spherical particle glued to the cantilever and the sample is usually a flat surface. Deflection of the cantilever is measured by a photodiode. As the sample is moved towards the sphere there is initially no movement of the cantilever (zero interaction force), but on closer approach the cantilever begins to deflect until eventually the two surfaces come into contact whereupon the cantilever deflects in compliance with sample movement. This final stage provides the zero for the separation distance and also serves to calibrate the instrument (Ducker *et al.*, 1991).

## 9.10 Solvation effects in colloid interactions

### 9.10.1 Solvent structuring in particle interactions

In the DLVO theory the solvent is treated as a continuum fluid having no structure. This approximation works well in many situations, but when the distance between the two solid surfaces is small it becomes necessary to consider the molecular structure of the solvent.

Two effects need to be distinguished: the interaction, solvation, between the solid and the solvent molecules, and a purely geometric effect where the solid surface causes some layering of the solvent molecules close to it. When two solid surfaces approach one another the geometric effect can become much more marked. The geometric effect dominates at lyophobic surfaces and is less significant, but still present at lyophilic surfaces where the solvation effect dominates. Both effects lead to oscillations in the interaction force between the hard surfaces that diminish rapidly with separation.

### Geometric effect

In the simplest case the solvent molecules are hard spheres that do not interact with the solid surfaces. The interaction between two approaching surfaces begins to oscillate because of the packing constraints. When the solvent molecules can arrange themselves into ordered layers there is attraction between the surfaces but when the separation is not a multiple of the molecular diameter of the solvent the solvent is disordered and there is repulsion. These effects have been observed in measurements of the interaction force (Horn and Israelachvili, 1981) and their origin is illustrated in Figure 9.14.

However, if the molecules are flexible or non-spherical the layer structuring becomes blurred and force oscillations may not be seen.

For hydrophobic surfaces in water there is a strong tendency for the water molecules towards molecular arrangements that minimize the number of unsatisfied potential hydrogen bonds. Such arrangements have a lower entropy than bulk water so that when two hydrophobic surfaces approach the entropically unfavoured water is ejected and the two hydrophobic surfaces attract. Experimentally the effect has been observed at separations up to 6 nm.

### Solvation effect

If the hard surfaces and the solvent have a significant interaction there will tend to be a layer of solvent molecules covering the surfaces: the surfaces will be solvated (hydrated if the solvent is water). There will also be a tendency towards a layered structure further from the surface. Thus when two solvated surfaces approach, the overlap of their solvation zones causes disruption of the solvation layers and this is experienced as a repulsion (Israelachvili, 1992, p. 266). If the solvent molecules are roughly spherical and rigid the interaction force will oscillate with a periodicity close to the mean molecular diameter of the solvent molecule and will decay with a half-decay length of about 1.5 diameters.

For water as solvent there may be groups on the solid surface that can hydrogen bond to water or carry charges that interact with the charges on the water molecule (see Section 2.9.3). There will then be considerable energy required to displace these hydrating water molecules and allow the surfaces to come into contact. In other words, there will be a strong short-range repulsion between the surfaces and the force will not oscillate with distance.

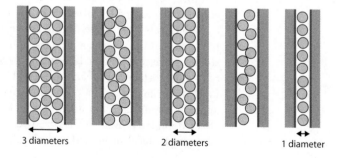

**Figure 9.14.** Packing of approximately spherical molecules of an inert non-polar liquid between two flat parallel surfaces (after Israelachvili, 1991, Fig. 13.2).

3 diameters    2 diameters    1 diameter

This effect provides explanations for the stability of hydrophilic colloids, the swelling of clays, the retention of a thin layer of solution in certain soap films, and the spacing between phospholipid bilayers.

### 9.10.2 Steric stabilization of colloids

The stabilization of emulsions by polymeric materials is discussed in Section 6.3.3 and similar considerations apply when the disperse phase is solid. Both natural and synthetic polymers may be used. Often the synthetic polymers are block or graft copolymers with hydrophilic segments and oleophilic segments. They will tend to adsorb on the surfaces of lyophobic colloidal particles with the lyophobic segments (the *anchor* moiety) on the surface and the lyophilic segments (the *stabilizing* moieties) out in the dispersion medium. It is the repulsive interaction of the lyophilic segments on approaching particles that prevents coagulation.

For a good stabilizing polymer the anchoring function must be sufficient to prevent displacement during a collision, the surfaces of the particles must be covered, and the loops and tails of the stabilizing moiety must extend far enough out into the medium.

There are several advantages of steric stabilization over stabilization by double-layer repulsion: it is effective in non-aqueous as well as aqueous dispersion media; it protects the colloid against coagulation by electrolytes; and it is effective over a wide range of colloid concentrations.

## 9.11 Self-assembled films

In this context, the term *self-assembly* describes the spontaneous deposition of a molecular layer from solution on to a solid substrate, usually by chemisorption. Thus one way the self-assembly process is distinguished from normal adsorption is by the strength of the interaction. The other distinguishing feature is that the molecules adsorb in an ordered fashion (assemble) usually as a result of van der Waals interactions between alkyl chains. Examples include the adsorption of alkanethiols on gold and some other metals, and alkyltrichloro-silanes on various hydroxylated surfaces (such as glass and silica). In most cases the reactive group on the adsorbate would be hydrophilic. Consequently, when the adsorbate is an amphiphile, as is often the case, the solvent should be an organic liquid to enable the oleophilic part of the amphiphile to extend more readily into the solvent phase.

The strong interaction means that all available adsorption sites will tend to be occupied and this usually leads to the formation of highly ordered (crystalline) adsorbed films known as self-assembled monolayers (SAMs). This ordering may be dictated by the arrangement of interacting groups of the solid surface, but in some cases it is the packing and interacting of neighbouring adsorbed molecules that is responsible.

$$X-(CH_2)_n-SiCl_3 + 3HO-Si^S \longrightarrow X-(CH_2)_n-Si-(O-Si^S)_3 + 3HCl$$

or

$$X-(CH_2)_n-SiCl_3 + HO-Si^S + 2H_2O$$
$$\longrightarrow X-(CH_2)_n-Si(OH)_2-(O-Si^S) + HCl$$

An important example is the change of a fully hydrophilic surface, such as clean glass or silica/silicon, to a surface that is not only hydrophobic but also oleophobic and autophobic (not wetted by the treating liquid). The treatment is simple: the substrate is dipped for an hour or two into a solution of alkyltrichlorosilane ($C_n$TS) in an organic solvent, rinsed with solvent, methanol, then water, and dried. The reaction is shown in Figure 9.15.

The adsorbed monolayer is not readily removed by chloroform (which is a good solvent for $C_n$TS) which suggests strong, probably covalent, bonding with the substrate (as in Figure 9.15) and in-plane hydrogen bonding with possibly some polymerization.

Tests of the stability of self-assembled films using washes with acid, alkali, and various organic solvents, and moderate heat (up to 130 °C) generally show that the stability is very good and appreciably better than that of the corresponding LB films. One exception is the attack of alkali on silicon-based materials where hydrolysis of the Si–O bond causes significant changes (Ulman, 1991, p. 254).

If the deposition of several layers is required, the exposed outer section of each adsorbed molecule would need to carry either a reactive group or a group that could be rendered reactive. However, to avoid polymerization within the solution phase it is essential that no interaction can occur between such terminal groups and the group reacting with the subphase. For example, if the X-group in Figure 9.15 is a vinyl group it can readily be altered (after completing adsorption of the silane to the silica surface) to another functional group which could then form the basis for a second chemisorbed layer.

Thus when deposition of the first monolayer involves reaction with a hydroxylated surface (as in Figure 9.15) and the terminal group of the adsorbed molecules is converted to hydroxyl, the second layer will be deposited by the same process as the first. Repetition of this procedure will build up a sequence of monolayers with all molecules oriented in the same direction, similar to Z-type LB deposition (see Section 5.4). Figure 9.16 gives an example where the silane reaction results in extensive cross-linking within each adsorbed monolayer as well as chemical bonding of each layer to the one beneath and chemical bonding to the substrate. Clearly this structure would be much more robust than a physically adsorbed LB film.

Such highly ordered films would be ideal for non-linear optics and molecular electronics applications, but there are difficulties, such as irregularities in the substrate surface and incomplete layer deposition, that have so far limited their potential. Furthermore, Ulman (1991) concludes his review

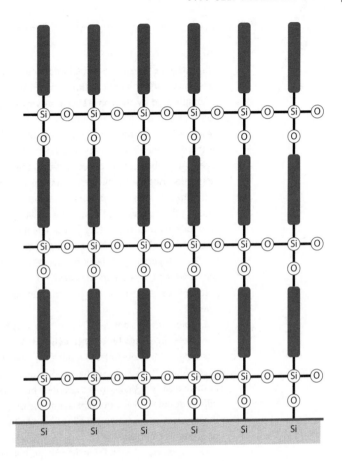

**Figure 9.16.** Schematic diagram of a self-assembled multilayer film. Blue sections indicate the alkyl chains. (After Ulman, 1991, p. 247.)

of this topic by pointing out that the polymeric nature of films formed from alkane-trichlorosilane means that they are not as ordered as films of alkane thiols on metals. It has also been argued (Sagiv, 1980) that polymerisation is only a problem when water is present. In strictly anhydrous conditions there will be no polymerization in the solution phase and only some possible polymerization within the adsorbed layer.

Alkyl thiols may be adsorbed on gold surfaces from a variety of solvents (both polar and non-polar). Longer-chain thiols (>12 carbons) give coherent, densely packed, usually tilted, monolayers that are stable in a range of environments. Importantly, they appear to be free of defects such as pin holes that could moderate the film properties (Porter *et al.*, 1987). Of particular interest are thiols with a functional group at the terminal end of the alkyl chain for, as we have seen, such groups can be designed to facilitate the attachment of second and subsequent layers. Simple terminal substituents (carboxylic acid, alcohol, amide, methyl ester) do not significantly alter the packing pattern (Nuzzo *et al.*, 1990). There are several other adsorbate-on-metal systems that have shown similar properties.

## SUMMARY

Adsorption at the solution–solid interface is usually measured by the change in solute concentration when the solution comes into contact with the solid. Adsorption of solute is complicated by the **competitive adsorption** of solvent molecules. Thus at very low concentrations, while the Langmuir or Freundlich isotherm equations may formally describe the adsorption they do not allow for the marked differences between solvents or the different effects of acyl chain length for adsorption on polar and non-polar surfaces. At higher solute concentrations the situation is described by the composite adsorption isotherm which takes solvent adsorption into account.

Charged surfaces in an aqueous solution attract ions of opposite charge and repel those of like charge setting up a pattern known as the **diffuse electrical double layer**. A modification of the theory of the electrical double layer is required when the ionic strength is high and ion size has to be considered. This leads to the concept known as the **Stern layer**: a single layer of ions at the surface of the solid, fairly firmly attracted to the solid and limited in concentration by ion size. Electrophoretic measurements measure the $\zeta$-potential which is considered to be the potential at the outside of this layer.

Colloidal dispersions are divided into **lyophilic colloids**, which form spontaneously and are stable, and **lyophobic colloids**, which are difficult to form and tend to be unstable. Added electrolytes reduce the stability of lyophobic colloids and may cause coagulation. The stability of lyophobic colloids is governed by the balance between the attractive van der Waals forces between particles and the repulsion that occurs when the diffuse double layers around each particle begin to overlap. The **DLVO theory** describes this interaction reasonably well for many situations, but there are other effects that it does not encompass.

One method of stabilizing lyophobic colloids is by the adsorption of amphiphilic polymers: the lyophobic sectors attach the polymer to the surface while the lyophilic sections project out into the solution and prevent the particles from approaching one another. This is known as **steric stabilization.**

In certain cases when there is very strong interaction between solute molecules and the solid surface the surface may become covered by a close-packed monolayer of solute. If the outer terminal groups of these solute molecules can be made reactive another layer may be added and the process repeated to generate a multilayer structure. Because the interactions between layers and with the solid are strong these **self-assembled films** have superior stability relative to LB films.

## FURTHER READING

Alexander, A. E. and Johnson, P. (1950). *Colloid Science*. Oxford University Press, London. A substantial and authoritative discussion of work up to publication.

Evans, D. F. and Wennerström, H. (1999). *The Colloidal Domain: Where Physics, Chemistry, Biology and Technology Meet. 2nd edn.* Wiley, New York.

Hamley, I. W. (2000). *Introduction to Soft Matter: Polymers, Colloids, Amphiphiles, and Liquid Crystals.* Wiley, New York.

Hunter, R. J. (1993). *Introduction to Modern Colloid Science*. Oxford University Press, Oxford.

Hunter, R. J. (2001). *Foundations of Colloid Science, 2nd edn*. Oxford University Press, Oxford.

Israelachvili, J. N. (1991). *Intermolecular and Surface Forces, 2nd edn*. Academic Press, London.

Napper, D. H. (1983). *Polymeric Stabilization of Colloidal Dispersions*. Academic Press, New York.

Ostwald, W. and Fischer, M. H. (1919). *Handbook of Colloid Chemistry, 2nd edn*. Blakiston's Son, Philadelphia. Of historical interest, written by two of the major figures in the early development of the subject.

Russel, W. B., Saville, D. A., and Schowalter, W. R. (1989). *Colloidal Dispersions*. Cambridge University Press, New York. An advanced treatment.

Sheludko, A. (1966). *Colloid Chemistry*. English edition. Elsevier, Amsterdam.

Ulman, A. (1991). *An Introduction to Ultrathin Organic Films from Langmuir–Blodgett to Self-Assembly*. Academic Press, San Diego. A thorough review of the entire subject with one section devoted to self-assembled films and also a discussion of applications.

Vold, M. J. and Vold, R. D. (1964). *Colloid Chemistry*. Reinhold, New York. A small paperback providing a useful introduction to the topic.

## REFERENCES

Bartel, F. E. and Fu, Y. (1929). *J. Phys. Chem.* **33**, 676.

Battersby, B. J., Sharp, J. C. W., Webb, R. I., and Barnes, G. T. (1994). *J. Microsc.* **176**, 110.

Ducker, W. A., Senden, T. J., and Pashley, R. M. (1991). *Nature* **353**, 239.

Hansen, R. S. and Craig, R. P. (1954). *J. Phys. Chem.* **58**, 211.

Horn, R. G. and Israelachvili, J. N. (1981). *J. Chem. Phys.* **75**, 1400.

Kipling, J. J. and Wright, E. H. M. (1962). *J. Chem. Soc.* **1962**, 855.

La Mer, V. K. and Barnes, M. D. (1946). *J. Colloid Sci.* **1**, 71.

La Mer, V. K. and Dinegar, R. H. (1950). *J. Amer. Chem. Soc.* **72**, 4847.

Nuzzo, R. G., Dubois, L. H., and Allara, D. I. (1990). *J. Amer. Chem. Soc.* **112**, 558.

Porter, M. D., Bright, T. B., Allara, D. L., and Chidsey, C. E. D. (1987). *J. Amer. Chem. Soc.* **109**, 3559.

Sagiv, J. (1980). *J. Amer. Chem. Soc.* **102**, 92.

Tiselius, A. (1937). *Trans. Faraday Soc.* **33**, 524.

Zeiser, E. M. and La Mer, V. K. (1948). *J. Colloid Sci.* **3**, 571.

## EXERCISES

**9.1.** Calculate and compare curves for the double layer interaction energy or the relative energy as a function of $\kappa s$ using the two approximations of Eq. (9.23).

**9.2.** Identical spherical particles of radius 80 nm are dispersed in an aqueous medium containing sodium chloride at a concentration of 3 mmol dm$^{-3}$. The temperature is 298 K. The effective Hamaker constant is $1 \times 10^{-19}$ J and the dielectric constant of the medium is 80.10. The permittivity of free space is $8.85 \times 10^{-12}$ F m$^{-1}$. Calculate the Debye length for the diffuse double layer around each particle. Electrophoretic measurements give a value of 45 mV for the $\zeta$-potential. Calculate the energy of interaction of two particles at a separation of 10 nm. Do these particles experience a mutual repulsion or attraction at this distance?

# 10 Biological interfaces

## 10.1 Introduction

Most of the interfaces that we have looked at so far have been formed by the meeting of two bulk phases without any physical barrier between them. However, when we look at interfaces in biological systems, plants and animals, we find that in many cases the bulk phases are separated by a thin film known as a membrane. This should not be too surprising as most of the liquid bulk phases are aqueous, so that without a membrane to separate them the liquids and their contents would mix indiscriminately.

Cell membranes are clearly some of the most important examples, without which many of the complex biological reactions that occur in cells could not occur. The cell membrane has two important functions: one is to contain the ingredients necessary for the cell to function; the other is to interact with the environment by allowing, or even helping, unwanted materials to leave the cell and needed materials to enter. Thus the permeability of membranes is a vitally important property.

There are other aspects of biology that involve interfaces. The fluid lining of the lung contains materials, known collectively as **lung surfactant**, without which breathing would be impossible, and the lack of lung surfactant is a significant cause of death in premature infants.

## 10.2 Membrane materials

Certain substances commonly occur as major components of biological membranes and interfaces so a preliminary review of their properties, particularly their surface properties, is desirable.

### 10.2.1 Phospholipids

Phospholipids are twin-chain surfactant molecules that form the basic matrix of most animal cell membranes, and are also important components of other biological surfaces such as lung surfactant. As lecithin, they are a well known food additive.

**Molecular structure**

Glycerol forms the central part of all phospholipids. To it are attached two long-chain fatty acids and a phosphate group which usually carries a nitrogen containing group such as choline or ethanolamine (Figure 10.1). The molecule is therefore amphiphilic and often the polar group is zwitterionic. The fatty acids in naturally occurring phospholipids usually have an even number of carbon atoms and may be saturated or unsaturated.

Other head groups encountered in nature include phosphatidyl glycerol (PG) and phosphatidyl serine (PS), where the groups attached to the phosphate are respectively $-CH_2-CH(OH)-CH_2OH$ and $-CH_2-CH(NH_3^+)-COO^-$ (both giving, with the negative phosphate, an anionic head group).

The acyl chains of membrane phospholipids have even numbers of carbon atoms with 16 and 18 carbon atoms dominating. Unsaturation is mostly in the *cis* conformation and, where present, is generally found in the acyl chain attached to the central carbon of the glycerol. Most phospholipids in membranes have one saturated acyl chain and one unsaturated as this ensures that the chain melting temperature is lower than physiological temperatures.

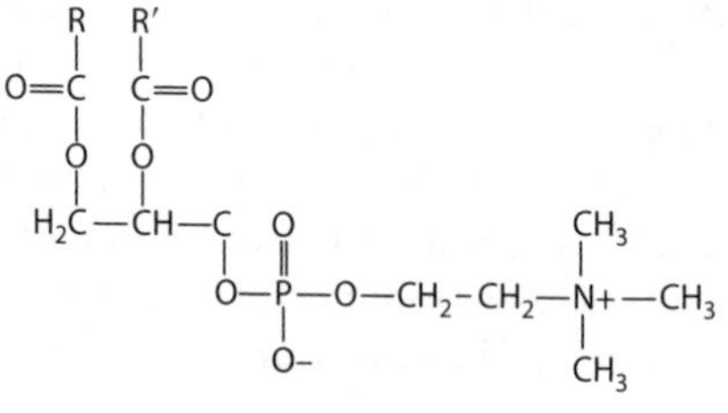

**Figure 10.1.** The structures of two common phospholipids at neutral pH values: R and R′ represent the alkyl chains of the fatty acids.

Phosphatidyl cholines (PC)

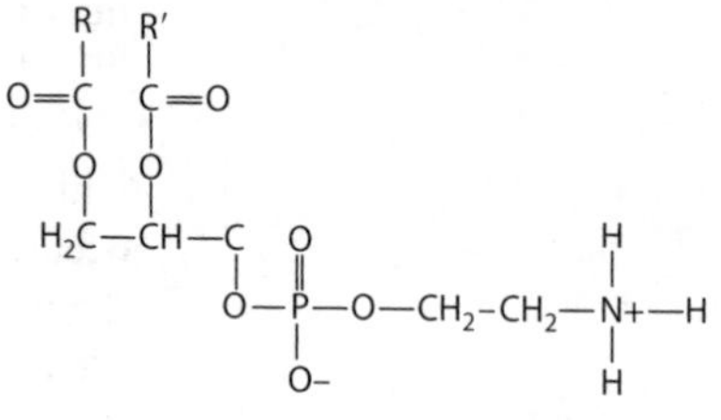

Phosphatidyl ethanolamines (PE)

In crystals, the phospholipids are arranged as stacks of bilayers, where the bilayers are tail-to-tail layers of close-packed molecules stacked much like the Y-type LB films described earlier (Section 5.4). The planes of the C-C-C zigzags of chains on the same molecule are mutually perpendicular. Generally the chains are tilted relative to the normal of the plane of the head groups. The orientations of the head groups of the PC, PE and PG lipids are preferentially parallel to the plane of the head groups.

### Monolayer characteristics

There are numerous reports on the monolayer properties of the phospholipids that occur more commonly in nature. The surface pressure–area isotherms depend on various factors: length of the acyl chains; unsaturation in the acyl chains; type of group attached to the phosphate; pH of the subphase; and temperature. We will only present here an outline of the properties of one of the more common phospholipids, dipalmitoyl phosphatidylcholine or DPPC, and indicate how changes in the above conditions might affect the monolayer properties.

The most prominent feature in the isotherms shown in Figure 10.2 is the flatter region seen at a surface pressure between 10 and $20\,\mathrm{mN\,m^{-1}}$. Although, during compression, the surface pressure rises continuously throughout this feature the process is nevertheless interpreted as a first-order phase transition between the liquid-expanded (Le) state and a condensed state (sometimes referred to as the *liquid crystal* to *gel* transition) (see Section 5.6.1). Evidence proving this interpretation is provided by Brewster angle microscopy supported by fluorescence microscopy which both show the coexistence of two monolayer phases throughout this region. There is also evidence to show that the gradually rising surface pressure can be attributed to minute amounts of contaminants in the monolayer. In expansion, this phase transition is often considered to be a chain-melting transition (Janiak *et al.*, 1976, 1979).

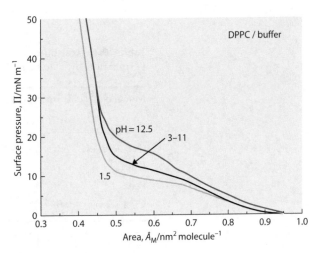

**Figure 10.2.** Surface pressure–area isotherms for DPPC monolayers on subphases of various pH values. (Data of Gorwyn and Barnes, 1990.)

At intermediate pH values the phosphatidylcholine head group is in the zwitterionic form and the isotherms do not show any change from pH 3 to 11. At lower pH values the addition of a proton to the phosphate group is a possibility which would lead to a net charge on the phospholipid and would therefore be expected to raise the surface pressure, but instead the experimental data show a small decrease. The most likely explanation of this result is the suggestion that the $pK_a$ value for the phosphate is reduced by the high concentration of charges at the surface compared with single ions in solution (Standish and Pethica, 1968). Another possibility is a change in the head group conformation or orientation.

At high pH values there is a small increase in surface pressure, but as the quaternary ammonium ion is unable to dissociate there can be no alteration of the net molecular charge. Again no satisfactory explanation for the surface pressure increase is available. With the phosphatidylethanolamines, on the other hand, the quaternary ammonium group is able to lose a proton at high pH values giving the molecule a net negative charge and generating surface pressures that are notably higher than those at low and intermediate pH values.

The main effect of small changes in the length of the acyl chains is seen mainly in the surface pressure for the expanded to condensed phase transition: shortening the chains raises the transition pressure, and lengthening them lowers it. Chain unsaturation or branching and increasing temperature would be expected to lead to higher surface pressures or larger monolayer areas.

Water permeation through some lipid monolayers has been measured by VanderVeen and Barnes (1985). The results are reported as permeation resistances (Eq. 5.18) and selected values are shown in Table 10.1. It is notable that the resistances of the phospholipids and cholesterol are significantly lower than the resistances of octadecanol. Both of these phospholipids have fully saturated acyl chains and even lower resistances would be expected for lipids with some unsaturation in one of the chains.

**Table 10.1.** Resistances of various monolayers to permeation by water at 25 °C.

| Monolayer | Permeation resistance, $r/s$ $cm^{-1}$ | |
| --- | --- | --- |
| | At $\varPi = 20\,mN\,m^{-1}$ | At $\varPi = 40\,mN\,m^{-1}$ |
| Octadecanol | 2.5 | 3.8 |
| DPPC | 0.13 | 0.35 |
| DSPC* | 0.38 | 1.4 |
| Cholesterol | 0.08 | 0.15 |

* DSPC is distearyl phosphatidylcholine.

## 10.2.2 Cholesterol

Cholesterol is the most common of the family of sterols, a group of alcohols with the same basic ring structure but differing in the side chain and peripheral groups.

**Molecular structure**

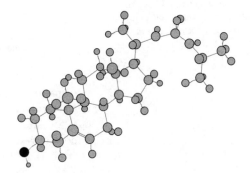

**Figure 10.3.** The molecular structure of cholesterol. Carbon atoms are shown as dark grey, hydrogen blue, and oxygen black.

**Monolayer characteristics**

The surface pressure–area isotherm of a cholesterol monolayer is very simple. On compression there is no appreciable rise in surface pressure until an area of about $0.40\,nm^2\,molecule^{-1}$ is reached whereupon there is a steep, practically linear, rise up to collapse. The areas involved suggest a close-packed monolayer structure with the flat ring structure of the molecules almost vertical. It is therefore surprising that GIXD scans (see Section 5.5.4) of such monolayers show no diffraction peaks, suggesting that the structure is disordered.

## 10.2.3 Phospholipid + cholesterol

**Monolayer characteristics**

There have been many reports that mixtures of cholesterol with phospholipids have monolayer areas that are considerably lower than the calculated sum of the areas of the individual components (see Eq. 5.15). This observation has generated considerable discussion with interpretations ranging from strong interaction between the components to (somewhat improbably) immiscibility. As a broad generalization, monolayers with more than 50 mol% cholesterol have characteristics similar to those of pure cholesterol.

## 10.2.4 Membrane proteins

Proteins that are located in membranes must necessarily have regions that are hydrophobic and must be water insoluble. This makes purification and crystallization difficult, if not impossible, and so limits the structural

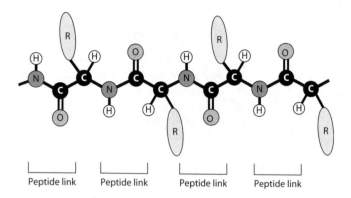

**Figure 10.4.** Basic structure of the protein chain. 'R' stands for the parts of the amino acid molecules not included in the chain structure.

Peptide link    Peptide link    Peptide link    Peptide link

information that may be obtained by such conventional techniques as X-ray diffraction and solution nuclear magnetic resonance. However newer methods for studying lipid-embedded proteins are now beginning to yield important results. Some of these results are outlined in this chapter.

Proteins are formed from amino acids joined by *peptide linkages*.

Shorter chains (molar mass $< 10\,\mathrm{kg\,mol^{-1}}$) are known as peptides or polypeptides. Chain flexibility arises from the possibility of rotation about the C–C bond and in proteins can result in complex and important conformations.

The sequence of amino acids is known as the *primary structure* while the *secondary structure* is a consequence of folding of the chain and hydrogen bonding between the C=O and N–H groups. The major structural elements found in proteins are α-helices and β-sheets, while intra- and intermolecular disulfide bonds are also important in determining the overall folding of the molecule.

Of major significance in the conformation of proteins is the hydrophobicity and hydrophilicity of various segments of the protein molecule. In water the protein molecule will tend to adopt a conformation in which the hydrophobic parts are sequestered away from the water while the hydrophilic parts are in contact with water. In this regard, the peptide linkages are essentially hydrophobic and the intramolecular hydrogen bonding mentioned above reduces the opportunity for the N–H and C=O groups to interact with water. Consequently, the hydrophobic or hydrophilic character of a segment of the protein molecule depends primarily on the nature of the constituent amino acids in that segment ('R' in Figure 10.4.).

## 10.3 Bilayers

### 10.3.1 Structure

In water, bilayers formed from phospholipids have the same basic structure as surfactants discussed earlier, with the headgroups on the outside in contact

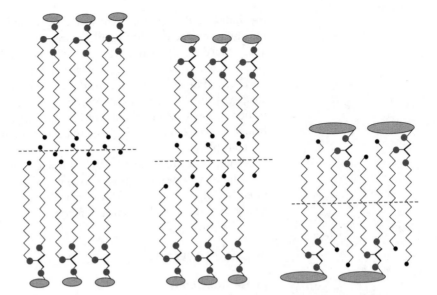

**Figure 10.5.** Schematic representation of some chain interdigitation patterns in bilayers in the gel state with untilted acyl chains. Acyl chain lengths equal on left and right, unequal in centre; large head groups on right; approximate bilayer mid-planes shown as dashed lines.

with the water and the acyl chains in the centre. There is, however, one important difference: each phospholipid molecule has two acyl chains. Even when these two chains are identical, the tilted orientation of the glycerol group to which they are attached means that one chain extends further from the phosphate group than the other (by about 1.3 methylene groups), and this effect may be larger with the many phospholipids where the two chains are different. When such molecules form a condensed monolayer or one half of a bilayer in the gel state, the outer surface would be irregular if the chains were oriented perpendicular to the plane of the head groups. With a floating monolayer this would not matter, but in a bilayer the chains must pack without leaving voids in the acyl chain layer. There are two possibilities: tilting of the molecules relative to the surface normal, and interdigitation of the chains from the two sides of the bilayer. Interdigitation is said to occur when some of the acyl chains extend beyond the mid-plane of the bilayer (Figure 10.5). X-ray crystallographic studies show that both tilting and interdigitation occur.

Above the gel $\leftrightarrow$ liquid-crystal transition temperature the chains are disordered, with the almost universal *trans* conformation of the acyl chains in the gel state replaced by chains with numerous *gauche* bonds. Thus the central hydrocarbon section of the bilayer would resemble a liquid hydrocarbon.

It is important to recognize that, in water, the bilayer structure is energetically very favourable: the hydrophilic head groups are in contact with water and the hydrophobic acyl chains are shielded from the water. There are approximately 11 water molecules associated with each phospholipid molecule in a bilayer (Jain, 1980, p. 70). Following Israelachvili (Section 6.7.3), the formation of a more-or-less flat bilayer would require a packing factor (Eq. 6.19) of about 1 (approximately a cylindrical molecular shape). In DPPC, for example, the cross-sectional area of the head group is very nearly the same as that of the two acyl chains, and spontaneous aggregation into bilayer structures has been reported to occur at concentrations below $10^{-10}$ mol dm$^{-3}$ (Jain and Wagner, 1980, p. 70). It is convenient to refer to such concentrations as a cmc even when it is a bilayer that forms rather than a micelle, although the term *critical bilayer concentration* or *cbc* has been suggested. It is important to note, however, that even though the concentration of single phospholipid molecules in water is extremely low (generally below $10^{-7}$ mol dm$^{-3}$) there is significant spontaneous transfer through the aqueous phase.

We note that the exposed edges of a flat bilayer would produce energetically unfavourable interactions between the hydrocarbon layer and water, so often, when we refer to a bilayer, it is not a perfectly flat sheet but curves gently so that a closed structure is formed.

### 10.3.2 Properties

#### Phase transitions

Just as with the phospholipid monolayers, the bilayers also exhibit the gel to liquid crystal or chain-melting transition. This main transition is accompanied by an increase in hydration of the bilayer surface. Thus the transition is affected by hydration and, like the monolayers, by temperature. Other transitions are also possible and with DPPC, for example, three low-temperature bilayer phases have been identified with transitions at temperatures below that of the main transition (Jain and Wagner, 1988). A ripple phase is often seen between the gel and liquid crystal phases (Figure 10.6). Some of these transitions are inhibited if the water content of the sample is limited. Normally in a living organism, there is sufficient water for complete hydration, but if dehydration through desiccation or freezing occurs there could be serious damage to cell membranes.

The saturation or unsaturation of the acyl chains has a major effect on the transition temperatures. With DPPC, for example, the main transition is at 41 °C, but when there is unsaturation present the transition is below 0 °C.

Isothermal phase transitions can be induced by changes in the concentration or composition of the solvent (Cevc and Kornyshev, 1993).

#### Interactions with cholesterol and proteins

The presence of cholesterol has a marked effect on the properties of phospholipid bilayers. The effect appears to be a reduction in the fluidity of the acyl chains in the liquid-crystal state and an increase in their fluidity in the gel state. Thus, in line with the monolayer observations, the chain-melting transition becomes less prominent and may be suppressed entirely (Yeagle, 1991, p. 125). However, the bilayer structure is destroyed if more than a limited amount of cholesterol is present.

The interaction of phospholipid bilayers with naturally occurring proteins generally leads to expansion with the protein entering partly or extensively into the hydrophobic core of the bilayer. Proteins that are mainly hydrophilic interact by electrostatic and hydrophobic forces and are positioned partly in

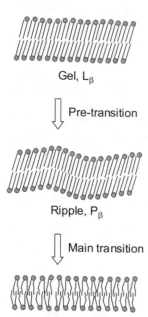

Gel, L$_\beta$

⟱ Pre-transition

Ripple, P$_\beta$

⟱ Main transition

Liquid crystal, L$_\alpha$

**Figure 10.6.** The gel (L$_\beta$) to liquid crystal (L$_\alpha$) phase transition, showing the intermediate ripple (P$_\beta$) phase.

the hydrophobic core of the bilayer and decrease the temperature of the phase transition. Essentially hydrophobic proteins penetrate deeply into or through the hydrophobic core of the bilayer and have little effect on the phase transition. Both types expand phospholipid monolayers and increase the permeability of vesicles, but these effects are much more marked with hydrophobic proteins (Yeagle, 1991).

Where the bilayer is a mixture of phospholipids the more hydrophobic components tend to be grouped near hydrophobic proteins.

## 10.4 Vesicles and liposomes

### 10.4.1 Phospholipid self-assembly

One of the intriguing features of phospholipids is their ability to self-assemble into a variety of interesting and potentially useful structures. Phospholipids generally have low solubility in water, but with the encouragement of ultrasonic agitation often form suspensions in which the phospholipid molecules are organized into particular structures depending on the temperature and the phospholipid mix. Often these structures can be observed in the electron microscope.

As mentioned above, the bilayer structure formed by phospholipids satisfies the amphiphilic property of the molecules except at the edges of a bilayer sheet. There is, therefore, a tendency for these sheets to bend and form closed structures that eliminate such edges. The closed sac-like structures known as vesicles or liposomes are a result. The term *vesicle* is a general term for a closed sac-like structure formed from amphiphilic bilayers. When the amphiphile is a lipid the term *liposome* may be applied. However, for this to occur the packing factor should be between 0.5 and 1 (Israelachvili, 1985, p. 255).

#### Structure and preparation

The vesicles produced from phospholipid bilayers are usually described by their size and the number of bilayers. Thus there are small (SUV) and large (LUV) unilamellar vesicles (both also described as ULV), and multilamellar vesicles (MLV) (Figure 10.7).

As suggested by the diagrams, phospholipid vesicles are formed at temperatures above the main transition temperature of the phospholipid. With the appropriate phospholipids, such as phosphatidyl cholines and ethanolamines in their zwitterionic forms, the formation of MLVs is easy: they often form spontaneously by placing the dry phospholipid in water (e.g. Figure 10.8). Cholesterol is readily incorporated into such vesicles. For ULVs, ultrasound may be used to break up MLVs. A narrow size distribution of ULVs can be prepared by extruding MLVs through a nucleopore polycarbonate membrane. These membranes have uniform pore sizes and the size chosen determines the size of the ULVs.

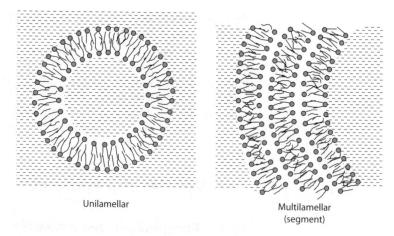

**Figure 10.7.** Schematic diagrams of phospholipid vesicles. The size of the molecules has been greatly exaggerated relative to the size of the vesicles.

Unilamellar

Multilamellar
(segment)

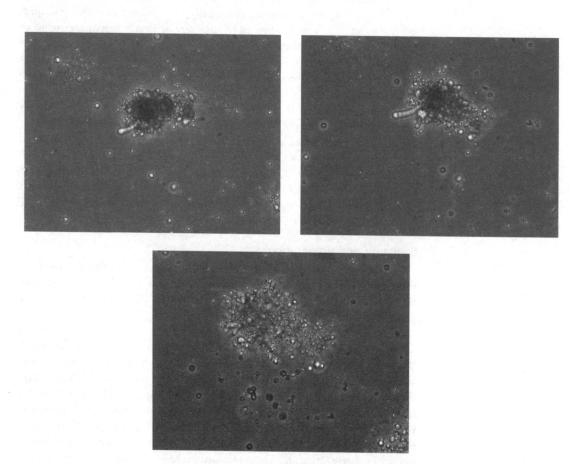

**Figure 10.8.** Dimyristoyl phosphatidylcholine in water above the main transition temperature showing (left to right) the spontaneous development of myelinic figures (finger-like protrusions from hydrated clusters of lipid) that will break up into multilamellar vesicles. The figures also show the general swelling of the cluster. (From B. J. Battersby.)

## 10.4.2 Electron microscopy

Electron microscopes operate at high vacuum and therefore require the sample to be stable at very low pressures. In particular, aqueous specimens dry instantly when exposed to this vacuum. For this reason the structures formed in phospholipid suspensions have, over many years, been observed in the electron microscope by staining with heavy ions such as phosphotungstic acid and uranyl acetate. There was always some question about whether these stains affected the structures that were being observed, but it was only with the development of newer techniques that did not involve staining that it was possible to show that indeed the stains did affect the observed structures (Talmon, 1983). These two stains were also shown to have profound effects on the surface pressure–area isotherms of several phospholipids (Gorwyn and Barnes, 1990).

An alternative to staining is cryoelectron microscopy which involves freezing the sample. In the freeze-fracture technique the sample is frozen, sheared to cause a fracture plane through the specimen, and a cast replica made for examination in the electron microscope. However, crystallization of water causes expansion and may disrupt the structures under examination so the freeze-fracture technique is open to criticism.

It is preferable to form ice that is amorphous rather than crystalline and this can be done by extremely rapid cooling to low temperatures, essentially bypassing the crystalline phases. Briefly, the technique is to deposit a drop of the test liquid on an electron microscope grid, blot off excess liquid, then plunge the grid rapidly into liquid propane cooled by liquid nitrogen. Later the grid can be transferred to and stored in liquid nitrogen before examination on the cold stage of a cryo-electron microscope. Liquid propane is used for the initial cooling rather than liquid nitrogen because liquid nitrogen forms a film of vapour over the sample which retards the cooling, allowing ice crystals to form. A device which enables the phospholipid suspension to be equilibrated at a selected temperature before transfer to an electron microscope grid has been described by Battersby et al. (1994).

## 10.4.3 Applications

### Model membranes

As vesicles can be prepared from natural constituents they closely resemble the lipid parts of natural membranes and are therefore useful models for cell membranes. However, while vesicles may replicate some of the functions and properties of natural membranes there are also important differences in structure and composition that affect their suitability as model membranes.

### Drug delivery

The structure of vesicles allows them to accommodate water-soluble and lipid soluble materials. This enables them to be used as delivery agents for drugs and also for cosmetics. In particular, the similarity of phospholipid vesicles to the bilayer of biological membranes suggests that they should be biocompatible.

For drug delivery the drug is solubilized in the liposomes which are injected into the blood stream where they carry the drug to the target while shielding it from premature release. Liposomes attach themselves to cell walls and are thus able to deliver drugs directly to cells. To improve the selectivity of these systems the liposomes may be modified by the addition of proteins selected to act as antibodies which bind with the target cells and in so doing disrupt the liposomes, releasing the drug. When the target is an area of inflammation, it is possible to select lipids that form liposomes that are stable at normal physiological temperatures but break down at the higher temperatures at the inflammation site. Vesicles sensitive to pH may also be used to target certain sites.

## 10.5 Cell membranes

### 10.5.1 Composition

The composition of biological membranes, sometimes called the plasma, varies greatly, but some general characteristics can be discerned. There are significant quantities of phospholipids with phosphatidylcholines and phosphatidylethanolamines predominating; cholesterol varies greatly, but can be up to half the total weight of phospholipids; and there can be up to five times as much protein as phospholipid (by weight). For a more detailed breakdown, see Yeagle (1991, p. 35).

There is a wide variety of cell types and membranes found in living organisms and a detailed discussion is outside the scope of this book. Further information can be found in Yeagle (1991). Here we will outline the general features.

### 10.5.2 Structure

The proteins associated with cell membranes can have a variety of forms and functions. In some cells there is a *membrane skeleton* comprising a network of protein filaments that lies inside the cell immediately beneath the cell plasma membrane. Particularly in the erythrocyte (red blood corpuscle) this skeleton helps to preserve the integrity of the cell against severe mechanical stress as it is pumped around the body.

Generally the phospholipids in membranes have 16 or 18 carbon atoms in each chain, and the chain attached to the central glycerol carbon is unsaturated. The cmc for the bilayer is extremely low and the chain melting transition is below $0\,°C$.

The generally accepted form of cell membranes is the **fluid-mosaic model** proposed in 1972 by Singer and Nicolson (1972) to describe what they termed *functional membranes*. This term excludes some more rigid membrane structures. The essential features of the fluid-mosaic model are illustrated in Figure 10.9 and described as follows.

- The proteins in intact membranes (*integral* proteins) are globular in shape rather than being spread as monolayers and they show appreciable amounts of α-helical structure.

- The phospholipids are present as a bilayer with the molecules in the liquid-crystal phase.

- The phospholipid bilayer forms the matrix of the membrane.

- The globular proteins are amphiphilic and interact with the appropriate section of the lipid bilayer: hydrophobic segments with the hydrophobic centre of the bilayer, hydrophilic segments with the hydrophilic head groups and with the surrounding water.

- Some proteins may extend completely through the bilayer, with hydrophilic segments at both ends.

- Interaction between proteins and phospholipids may, in some cases, affect the activity of the protein. On the other hand the properties of the phospholipids are little altered, suggesting that the interactions are restricted to the small proportion of lipids in close contact with protein.

- Lateral movement within the membrane is possible and there is some evidence that the more hydrophobic lipids in a mixed lipid bilayer are preferentially attracted to amphiphilic proteins in the membrane.

- Proteins that are not an integral part of the membrane (*peripheral* proteins) may have a specific interaction with the exposed segments of an integral protein.

- Two or more integral proteins may interact within the membrane to form a specific aggregate without altering the mosaic structure of the membrane.

There have been refinements of the fluid-mosaic model since its inception, but the basic concept has remained. Moreover, the structures are now being explored by new techniques that are capable of providing structural information about proteins embedded in lipid bilayers (Torres *et al.*, 2003).

For example, the technique of electron crystallography uses electron microscopy to investigate membrane proteins that have formed two-dimensional

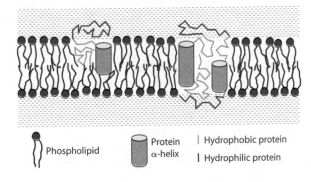

**Figure 10.9** Schematic cross-section of the fluid-mosaic model membrane. (After Singer and Nicolson, 1972.)

Phospholipid   Protein α-helix   | Hydrophobic protein   | Hydrophilic protein

planar ordered structures. The difficulty of forming three-dimensional crystals is thereby avoided. Both the imaging and the diffraction modes of the electron microscope are combined so that the 'phase problem' of conventional X-ray crystallography does not arise. Atomic force microscopy and various forms of nuclear magnetic resonance spectroscopy are also contributing important data. Attention is being directed particularly towards determining the orientation and packing of the α-helices in membranes.

### 10.5.3 Function: selective permeability

As indicated earlier there are two principal functions for biological membranes: retaining cell contents and controlling the transport of materials into and out of the cell. Both functions are essentially described by the term selective permeability.

Some elements of the transport of molecules through interfaces and membranes have been described earlier: the permeability of soap films to gases was treated in Chapter 4; in Chapter 5 we discussed the evaporation of water through floating monolayers; and in Chapter 6 liquid–liquid extraction, osmosis and semipermeable membranes, dialysis, and the Donnan equilibrium were treated. All of the concepts described in these discussions are relevant to the transport of molecules through biological membranes, but the situation with biological systems is much more complex. Not surprisingly, there has been an enormous amount of research on this topic and the literature is immense. Consequently we are unable to include more than a very brief outline of the topic and leave more detailed treatment to books in the Further Reading list.

Transport of substances through a membrane can involve either a solubility and diffusion mechanism or some form of facilitated transport through particular sites in the membrane.

Treatment of the solubility and diffusion mechanism is similar to the treatment of transport through soap films in Section 4.9.3 and will not be discussed further here. The driving force for such movement is the concentration difference on the two sides of the membrane, but it is worth noting that water, even though its concentration is the same on both sides, does move across biomembranes. This is demonstrated by the exchange of radio-labelled water which eventually leads to equal concentrations of labelled water on both sides. In this connection we recall that the evaporation rate of water through spread monolayers is reduced to a much smaller extent by phospholipid monolayers than by long-chain alcohol monolayers (see Table 10.1). The possibility of osmotic flow also has to be considered when there are differences in solute concentration on the two sides of the membrane (Section 6.7.1). One further point to note is that the solubility and diffusion of non-electrolytes in hydrocarbon solvents may differ significantly from their values in the relatively structured hydrocarbon region of a biomembrane.

As facilitated transport occurs at specific sites in the membrane the rate may reach a maximum due to saturation of the sites or may be reduced because some sites are blocked by inhibitor molecules. Transport mechanisms can be of two kinds: a **carrier** mechanism where the permeant binds to a site on the membrane which then moves to the other face of the membrane where the permeant is released; and a **channel** or **pore** mechanism where there is a more-or-less permanent opening through which the permeant can pass. Carrier processes may be highly specific, but channels are much less specific.

For the **carrier transport** of ions the selectivity of the process depends on the relative interaction free energies of the ions with their hydration shells and with the binding site of the carrier. Thus, for example, when the binding energy of the carrier site for alkali cations is small, the selectivity will depend on the hydration energies and we then have the familiar lyotropic or Hofmeister sequence (Section 9.9.2): $Cs^+ > Rb^+ > K^+ > Na^+ > Li^+$, where $Li^+$, the smallest ion, is the most strongly hydrated and the least readily transported. Of particular interest here is the difference between sodium and potassium ions where the hydration free energies are respectively $-395$ and $-314 \, kJ \, mol^{-1}$, meaning that sodium is transported less readily than potassium. Conversely when the binding energy between carrier site and cations is strong the hydration energies are less relevant and it is the closeness of approach of the bare cation to the carrier that determines the selectivity. Thus the sequence of alkali metal cations given above is now reversed: $Li^+$ is the most favoured.

These processes are further complicated by electrical potentials across membranes. Such potentials can arise from the movement of ions, ionization of surface groups, or Donnan equilibrium.

There is strong evidence for the transport of solutes through **aqueous pores** in biomembranes. Firstly, for small solutes the permeability is inversely related to their size rather than to their distribution coefficients between water and lipid (as required for the solubility and diffusion mechanism). Secondly, modification of the proteins can block the transport of small solutes. There thus appear to be water-filled pores with diameters between 0.3 and 0.8 nm in cell membranes. The transport of ions depends on the electrochemical gradient they experience as well as on their size.

Probably the most unusual concept associated with biomembranes is **active transport** where the solute moves against its gradient in electrochemical potential. Thus the free energy change for the transport process is positive so that a source of free energy is required. Various mechanisms for providing that have been identified.

- In co-transport the movement is coupled to the down-gradient transport of another solute. The formation of a complex between the two solutes is needed as a prerequisite.

- **Derivatization** of the solute during or after transport changes the concentration gradients and thus enables permeation to occur.

- In transport using ATP (adenosine triphosphate) the energy comes from the hydrolysis of ATP to ADP (adenosine diphosphate):

$$ATP + H_2O \longrightarrow ADP + phosphate$$

giving $\Delta G^{\theta} = -30\,\mathrm{kJ\,mol^{-1}}$. It is fortunate that in cells there is no catalyst that would allow this reaction to proceed to equilibrium, which strongly favours the right, and consequently the ATP is retained and is available to provide energy for activated transport (and other reactions).

- There are several other more specialized transport mechanisms where the required energy comes from reactions other than ATP hydrolysis. Details can be found in Jain and Wagner (1980).

## Example

Consider a cell in which the concentration of sodium ion is 20 times higher outside the cell than inside, the concentration of potassium ion is 20 times higher inside than outside, the potential difference across the membrane is 80 mV with the inside negative relative to outside, and the temperature is 37 °C. The differences in electrochemical potential are given by

$$\bar{\mu}_i^{\mathrm{out}} - \bar{\mu}_i^{\mathrm{in}} = RT\ln(c_i^{\mathrm{out}}/c_i^{\mathrm{in}}) + z_i F(\phi^{\mathrm{out}} - \phi^{\mathrm{in}}).$$

For sodium the difference in electrochemical potential is $15.4\,\mathrm{kJ\,mol^{-1}}$, but for potassium it is $0\,\mathrm{kJ\,mol^{-1}}$. Thus potassium is at equilibrium with this potential difference present, but sodium has a higher electrochemical potential outside than inside and so would tend to move from outside to inside. Active transport is therefore required to move sodium in the opposite direction and the minimum energy required to do this is $15.4\,\mathrm{kJ\,mol^{-1}}$. The hydrolysis of ATP provides ample energy to support this process.

A contributing factor to the electrical potential difference is the Donnan effect and the associated Donnan potential (see Section 6.7.1). Cells (phase $\alpha$) usually contain significant concentrations of macro-ions and if these carry negative charges the relationships developed in Section 6.7.1 indicate that

$$c_{\mathrm{Na}}^{\alpha} > c_{\mathrm{Na}}^{\beta} \quad \text{and} \quad c_{\mathrm{Cl}}^{\alpha} < c_{\mathrm{Cl}}^{\beta}$$

so, by Eq. (6.16)

$$\Phi^{\beta} > \Phi^{\alpha}.$$

The potential inside the cell is lower than that outside.

## Osmotic effects: tonicity

As biological cells contain solutes that are unable to pass through the cell membrane they have an osmotic pressure. If, for example, erythrocytes (red blood cells) are placed in water there is an osmotic flow of water into the cells which eventually causes them to rupture: a process called haemolysis.

However, the cell membranes are somewhat imperfect semipermeable membranes as they often permit the passage of certain solutes as well as of water. For example, the erythrocyte cell wall is permeable to solutes such as urea, boric acid, and ammonium chloride. It is therefore desirable to have a term other than osmosis to describe the behaviour of such membranes. That term is tonicity, but it has to be recognized that all cell membranes are not alike and that some solutes may be able to permeate one membrane but not another.

Tonicity can be measured by the haemolytic method: the effect of solutions of the relevant solute on the appearance of red blood cells. Isotonic solutions have no effect, as there is no net movement of water or other solutes across the membrane; hypotonic solutions allow water to permeate into the cells causing them to swell and burst; hypertonic solutions draw water out of the cells so that they shrink and become wrinkled.

### The sodium pump

As cells contain macromolecules with their associated counter ions they have an osmotic pressure which would tend to cause water to enter the cell and possibly cause rupture. It is therefore essential for the cell to avoid a build-up of ions inside the cell (by the Donnan effect, for example) and to try to maintain an adequate concentration of ions outside the cell. The sodium pump performs this function by pumping electrolyte from the cell to its surroundings.

The sodium pump operates by pumping sodium ions out of the cell, but this can only be effective if potassium ions do not move in the opposite direction to replace them. Thus the pump mechanism must discriminate between $Na^+$ and $K^+$ ions and an examination of the mechanisms discussed above indicates that a carrier mechanism might work. A variety of carriers have been found, but the details are complex and vary from system to system and thus lie outside the scope of this book. Details of a number of systems can be found in Yeagle (1991).

For the erythrocyte the equilibrium concentrations of the relevant ions are:

$$\text{Inside}: Na^+ = 0.3\,\text{mM}; \; K^+ = 10\,\text{mM}$$
$$\text{Outside}: Na^+ = 30\,\text{mM}; \; K^+ = 0.25\,\text{mM};$$
$$\text{ouabain (ATPase inhibitor)} \approx 1\,\mu M$$

## 10.5.4 Surfactant effects

Certain surfactants, particularly the alkyl trimethyl ammonium salts, appear to disrupt some biological membranes and are consequently used as antibacterial agents. It is thought that they disrupt the outer membranes of the bacteria.

We have previously mentioned (in Section 4.7.6) that impurities in a surfactant lower the surface tension and the apparent cmc. An understanding of this phenomenon enabled Alexander and Trim (1946) to explain the effects of various surfactants on the penetration of hexylresorcinol into the pig round worm (*Ascaris lumbricoides*). With a constant concentration of hexyl resorcinol, increasing the concentration of the surfactant (such as cetyl trimethyl ammonium bromide, sodium oleate, or sodium cholate) increased the rate of penetration to a maximum followed by a fall. Penetration was almost completely inhibited at high surfactant concentrations. Alexander and Trim recognized that the formation of micelles was probably the cause of the decline in activity. However, the cmc values of the pure surfactants did not coincide with the maxima in activity, but in the presence of hexylresorcinol there were significant changes in the interfacial tension curves with lower values at low surfactant concentrations and a pronounced minimum.

Reference to Figure 4.20 shows that the hexylresorcinol acts like an impurity in the surfactant: at low surfactant concentrations it lowers the surface tension and decreases the cmc, and as micelles form it is solubilized. In the presence of resorcinol the apparent cmc values occurred at the same concentrations as the penetration maxima. Thus the surfactants disrupt the membranes of the round worm enabling the resorcinol to penetrate more readily, but once micelles start to form the resorcinol is progressively locked up in the micelles.

However, it is worth noting that addition of surfactants may produce a very different effect when the biologically active material is a solid in suspension. O'Neill and Alexander (1963) observed that the effect of DDT (as a colloidal suspension or attached to a clay + lignin sulfonate suspension) on flour beetles decreased with increasing concentration of various surfactants up to the cmc. They attributed this effect to adsorption of surfactant on both DDT and beetle cuticle hindering the adhesion of DDT to the cuticle.

### 10.5.5 Molecular recognition

Molecular recognition is characterized as a strong, specific interaction between two molecules without covalent bonding. While it usually involves a solid surface that has been functionalized so that it interacts with a specific molecule in solution, floating monolayers may also be similarly functionalized.

A system that has been extensively investigated by Ringsdorf and associates (1991) is the interaction between biotin ($C_{10}H_{16}N_2O_3S$) and streptavidin, a protein. In one series of experiments, self-assembled films of biotinylated mercaptans and disulfides on gold were exposed to solutions of streptavidin and the amounts adsorbed measured. Surfaces with a high concentration of biotin bound less protein than those with a lower surface coverage indicating the importance of accessibility to the biotin.

## 10.6 Lung surfactant

Inside the alveoli of the lung is a thin film of liquid, which is highly curved as individual alveoli are very small. It is not surprising, then, that the alveoli have an inherent tendency to collapse brought about by the high surface tension at the air–liquid interface. Breathing is made possible by a phospholipid-rich film at the interface that maintains the surface tension at very low levels. In the absence of surfactant, the collapse of alveoli progresses to a condition known as respiratory distress syndrome (RDS), and unless breathing support is maintained the condition is frequently fatal. The condition is particularly prevalent among premature infants, with the majority of infants born at 28–30 weeks' gestation, and a number of low birth weight infants showing symptoms of RDS. Despite intensive research over many years, the remarkable action of natural lung surfactant (NLS, also known as pulmonary surfactant) is poorly understood. Nevertheless, the efforts of scientists have resulted in successful trials of artificial surfactant treatments for RDS in 1981, and in 1989 the US Federal Drug Administration approved two surfactant formulations for clinical use.

### 10.6.1 Natural lung surfactant: composition and function

NLS is a complex mixture of lipids and proteins. The bulk of the material (about 90%) is made up of phospholipids, with phosphatidylcholine dominating and in turn dipalmitoyl phosphatidylcholine (DPPC) making up the majority of the phosphatidylcholines. The other major components are proteins, making up about 10% of NLS. Four proteins have been found, known universally as SP-A (surfactant protein A), SP-B, SP-C, and SP-D. Of these, SP-A and SP-D are hydrophilic while the other two are very hydrophobic.

### 10.6.2 Lung surfactant production and life cycle

The important components of the lung surfactant process are:

- lamellar bodies, based on lipid bilayers, secreted by Type II cells into the alveolar subphase;
- tubular myelin, which is a highly ordered lattice-like structure containing lipids and proteins, and which is formed from multiple lamellar bodies;
- particles, either DPPC-rich or DPPC-poor, the latter which can be retrieved by the Type II cells for recycling;
- a surface associated phase;
- the monolayer phase at the air–liquid interface.

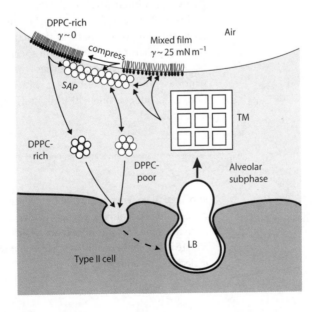

**Figure 10.10** Life cycle of lung surfactant in the alveolar subphase (redrawn after Goerke, 1998). LB: lamellar bodies; TM: tubular myelin; SAP: surface associated phase.

A commonly accepted version of the life cycle of lung surfactant is shown in Figure 10.10. Although details of the process are still debated, the basic steps in the production and transport of surfactant are thought to be as shown.

### 10.6.3 Lung surfactant surface activity

The functions of individual components of NLS have proven difficult to unravel, which is hardly surprising given the complex nature of the system, and the demanding circumstances under which it has to operate. Nevertheless, a great deal has been deduced from studies using film balances, captive bubble surfactometry, fluorescence microscopy and reflectivity methods among others. In order to function, the surfactant film at the alveolar air–water interface needs to lower the surface tension to less than $1\,mN\,m^{-1}$, and to be able to maintain a suitably condensed film under conditions of rapid expansion and contraction. It is generally believed that only a monolayer of DPPC can meet this requirement under physiological conditions. It is thought that the surfactant monolayer must undergo an enrichment process during the breathing cycle, either by excluding non-DPPC components or by insertion of DPPC into the monolayer. This implies that there is material under the monolayer, but associated with it, that acts as a reservoir for material in the refinement process. This multilayer structure is called a surface-associated phase.

The hydrophobic proteins, SP-B and SP-C, play a key role in modulating the arrangements of surfactant phospholipids in bilayer and monolayer structures, and are instrumental in the transfer of surface active materials between the different structural assemblies shown in Figure 10.10. SP-A and SP-D both have roles in the defense of the lungs from infection, while SP-A is also involved in the formation of the structures that transport lipids from the cells where they are produced to the air/water interface where they act, in particular tubular myelin.

Research into pulmonary surfactant continues with the goal of a fuller understanding of the natural surfactant, so that this knowledge will lead to the next generation of surfactant therapies for RDS. It is likely that these will be fully synthetic, using either recombinant proteins or fully synthetic peptides, and will therefore avoid the potential pitfalls of using animal-derived products, but will ultimately emulate the properties of this remarkable material.

## SUMMARY

In biological systems, many of the interfaces are between aqueous phases of different composition so a distinct physical barrier, a membrane, is required to keep them separate. Thus the contents of cells are retained by cell membranes, but another essential feature is that these membranes allow, and sometimes assist, the movement of certain molecules into and out of the cells.

The major components of biological membranes are phospholipids, cholesterol, and proteins. These are all amphiphilic materials and it is this feature that enables them to form membranes. The basic structure of a biological membrane is the bilayer: two phospholipid monolayers back-to-back with the hydrophobic parts of the phospholipids in the centre and the hydrophilic parts outwards interacting with the aqueous environment. Cholesterol and proteins are also incorporated in ways that conform with their amphiphilic properties and modify the membrane properties. For example, much of the movement of molecules and ions through the membrane is controlled by these incorporated proteins. This model is known as the **fluid-mosaic model membrane.**

The **selective permeability** of biological membranes can operate through a number of different mechanisms which are briefly discussed.

In the mammalian lung the inner lining is covered by a thin film of liquid. This film expands and contracts during breathing so it is essential that the surface tension of the air–liquid interface should be low. This is accomplished by the presence in this film of certain phospholipids and proteins in a mixture known as **lung surfactant.** This lung surfactant is only formed in the late stages of gestation so may be lacking in premature infants. In such case the infant has difficulty in breathing, suffering **respiratory distress syndrome**, RDS, and may die unless treated with an artificial lung surfactant.

## FURTHER READING

Cevc, G. and Marsh, D. (1987). *Phospholipid Bilayers: Physical Principles and Models, Vol 5.* Wiley-Interscience, New York.

Fendler, J. H. (1982). *Membrane Mimetic Chemistry: Characterizations and Applications of Micelles, Microemulsions, Monolayers, Bilayers, Vesicles Host-Guest Systems, and Polyions.* Wiley, New York.

Gregoriadis, G. (1993). *Liposome Technology: Liposome Preparation and Related Techniques*, Vol. 1, 2nd edn., CRC Press, Boca Raton.

Hamley, I. W. (2000). *Introduction to Soft Matter: Polymers, Colloids, Amphiphiles, and Liquid Crystals.* Wiley, Chichester. There are substantial chapters on the four topics in the title, together with some essential background material on surface science.

Jain, M. K. and Wagner, R. C. (1988). *Introduction to Biological Membranes*, 2nd edn. Wiley, New York. A well-presented treatment at the introductory level.

Yeagle, P. (ed.) (1991). *The Structure of Biological Membranes.* CRC Press, Boca Raton. A substantial and comprehensive volume with chapters ranging from the structure and properties of lipids and proteins, through membrane structure, to the transport properties of various membranes.

Exerowa, D. and Kruglyakov, P. (1998). *Foams and Foam Films. Theory, Experiment and Application.* Elsevier, Amsterdam. Includes a section on lung surfactant.

Dickinson, E. (1992). *An Introduction to Food Colloids.* Oxford University Press, Oxford. Also other similar titles by this author.

Various authors (1998). Special edition of *Biochimica Biophysica Acta*, vol. 1408. Contains a number of excellent articles on lung surfactant composition and action.

## REFERENCES

Alexander, A. E. and Trim, A. R. (1946). *Proc. Roy. Soc. London B* **133**, 220.

Battersby, B. J., Sharp, J. C. W., Webb, R. I., and Barnes, G. T. (1994). *J. Microsc.* **176**, 110.

Cevc, G. and Kornyshev, A. A. (1993). *J. Chem. Phys.* **98**, 5701.

Goerke, J. *Biochim. Biophys. Acta* (1998), **1408**, 79.

Gorwyn, D. and Barnes, G. T. (1990). *Langmuir* **6**, 222.

Häussling, L., Ringsdorf, H., Schmitt, F.-J., and Knoll, W. (1991). *Langmuir* **7**, 1837 and earlier references therein.

Jain, M. K. and Wagner, R. C. (1980). *Introduction to Biological Membranes*. Wiley, New York, Chapter 12.

Janiak, M. J., Small, D. M., and Shipley, G. G. (1976). *Biochemistry* **15**, 4575.

Janiak, M. J., Small, D. M., and Shipley, G. G. (1979). *J. Biol. Chem.* **254**, 6068.

O'Neill, D. K. and Alexander, A. E. (1963). *J. Sci. Food Agric.* **14**, 442.

Singer, S. J. and Nicolson, G. L. (1972). *Science* **175**, 720.

Standish, M. M. and Pethica, B. A. (1968). *Trans. Faraday Soc.* **64**, 1113.

Talmon, Y. (1983). *J. Colloid Interface Sci.* **93**, 366.

Torres, J., Stevens, T. J., and Samsó, M. (2003). *Trends in Biochemical Sciences* **28**, 137.

VanderVeen, R. J. and Barnes, G. T. (1985). *Thin Solid Films* **134**, 227.

# Index